T0321596

DIFFERENTIAL GEOMETRY IN ARRAY PROCESSING

Athanassios Manikas
Imperial College London, UK

DIFFERENTIAL GEOMETRY IN ARRAY PROCESSING

ICP

Imperial College Press

Published by

Imperial College Press
57 Shelton Street
Covent Garden
London WC2H 9HE

Distributed by

World Scientific Publishing Co. Pte. Ltd.
5 Toh Tuck Link, Singapore 596224
USA office: 27 Warren Street, Suite 401–402, Hackensack, NJ 07601
UK office: 57 Shelton Street, Covent Garden, London WC2H 9HE

British Library Cataloguing-in-Publication Data
A catalogue record for this book is available from the British Library.

DIFFERENTIAL GEOMETRY IN ARRAY PROCESSING

ISBN-13 978-1-86094-422-2
ISBN-10 1-86094-422-1
ISBN-13 978-1-86094-423-9 (pbk)
ISBN-10 1-86094-423-X (pbk)

Typeset by Stallion Press
E-mail: enquiries@stallionpress.com

Printed in Singapore

To my wife, Eleni

Preface

During the past few decades, there has been significant research into sensor array signal processing, culminating in the development of super-resolution array processing, which asymptotically exhibits infinite resolution capabilities.

Array processing has an enormous set of applications and has recently experienced an explosive interest due to the realization that arrays have a major role to play in the development of future communication systems, wireless computing, biomedicine (bio-array processing) and environmental monitoring.

However, the "heart" of any application is the structure of the employed array of sensors and this is completely characterized [1] by the *array manifold*. The *array manifold* is a fundamental concept and is defined as the locus of all the response vectors of the array over the feasible set of source/signal parameters. In view of the nature of the array manifold and its significance in the area of array processing and array communications, the role of differential geometry as the most particularly appropriate analysis tool, cannot be over-emphasized.

Differential geometry is a branch of mathematics concerned with the application of differential calculus for the investigation of the properties of geometric objects (curves, surfaces, etc.) referred to, collectively, as "manifolds". This is a vast subject area with numerous abstract definitions, theorems, notations and rigorous formal proofs [2,3] and is mainly confined to the investigation of the geometrical properties of manifolds in three-dimensional Euclidean space $\mathcal{R}^3$ and in *real* spaces of higher dimension.

However, the array manifolds are embedded not in real, but in N-dimensional complex space (where N is the number of sensors). Therefore, by extending the theoretical framework of $\mathcal{R}^3$ to complex spaces, the underlying and under-pinning objective of this book is to present a summary of those results of differential geometry which are exploitable and of practical interest in the study of linear, planar and three-dimensional array geometries.

Thanassis Manikas — London 2003
a.manikas@imperial.ac.uk
http://skynet.ee.imperial.ac.uk/manikas.html

Acknowledgments

This book is based on a number of publications (presented under a unified framework) which I had over the past few years with some of my former research students. These are Dr. J. Dacos, Dr. R. Karimi, Dr. C. Proukakis, Dr. N. Dowlut, Dr. A. Alexiou, Dr. V. Lefkadites and Dr. A. Sleiman. I wish to express my pleasure in having had the opportunity in working with them.

I am indebted to my research associates Naveendra and Jason Ng as well as to my MSc student Vincent Chan for reading, at various stages, the manuscript and for making constructive suggestions on how to improve the presentation material.

Finally, I should like to thank Dr. P. Wilkinson who had the patience to read the final version of the manuscript. His effort is greatly appreciated.

Keywords

linear arrays, non-linear arrays, planar arrays, array manifolds, differential geometry, array design, array ambiguities, array bounds, resolution and detection.

Contents

Chapter 1

Introduction

An array system is a collection of sensors (transducers) which are spatially distributed at judicious locations in the 3-dimensional real space, with a common reference point. How the sensors are spatially distributed (array geometry) is influential not only on the overall array capabilities but also on its "abnormalities." The type of the sensors varies with the application and sensors can take a wide variety of forms. Some common examples of sensors include electromagnetic devices (such as RF antennas, optical receivers, etc.) and acoustic transducers (such as hydrophones, geophones, ultrasound probes, etc.).

The signals at the array elements contain both temporal and spatial information about the array signal environment which is usually contaminated by background and sensor noise. Thus, the main aim of array processing is to extract and then exploit this spatio-temporal information to the fullest extent possible in order to provide estimates of the parameters of interest of the array signal environment. Depending on the application, typical parameters of interest associated with emitting sources (i.e. signals that use the same frequency and/or time-slot and/or code) can be the number of incident signals, Directions-of-Arrival (DOAs), Times-of-Arrival (TOAs), ranges, velocities etc. Indeed, with an array system operating in the presence of a number of emitting sources, and by observing the received array signal-vectors $\underline{x}(t)$, the following four general problems are of great interest:

(1) Detection problem — concerned with the determination/estimation of the number of incident signals. This problem is essentially the spatial analogue of model order selection in time-series analysis. Thus, the most popular methods for the solution of this problem are based on "Akaike Information Criterion" (AIC) [4] and the "Minimum Description Length" (MDL) criterion [5,6]. Both methods involve the

minimization of a function of the noise eigenvalues of the array output covariance matrix.

(2) Parameter estimation problem — where various signal and channel parameters are estimated. One important problem of this type is the "Direction Finding" (DF). In this case the parameters of interest are the bearings of emitters/targets (e.g. [7]). This problem is essentially the spatial analogue of the frequency estimation problem in time-series analysis.

(3) Interference cancellation (or reception problem) — the acquisition of one (desired) signal from a particular direction and the cancellation of unwanted co-channel interfering signals (or jammers), from all other directions. When the desired signal and the interference occupy the same frequency band, temporal filtering is inappropriate. However, the spatial separation of the sources can be exploited using an array of sensors (e.g. an antenna array). This operation falls, in array processing terms, under the general heading of "beamforming" while, in communication systems terms, a beamformer is a "linear receiver" (e.g. [8]).

(4) Imaging — here the parameters of interest are the shapes and sizes of various objects in the environment. These are typically determined by the generation of two- or three-dimensional maps depicting some feature of the received signals (e.g. intensity) as a function of their spatial coordinates (e.g. [9]).

The four types of problem described above are inter-related and the solution to one problem may result in a partial or complete solution to another. For example, the successful operation of all parametric parameter estimation algorithms requires solving firstly the detection problem (i.e. *a priori* knowledge of the number of emitters present). Furthermore, once the number and directions-of-arrival (DOAs) of signals received at the array site are estimated by solving the detection and direction-finding problems, nulls may be readily placed along the directions of the unwanted signals, hence achieving interference cancellation.

The applications of arrays in various scientific disciplines (such as the ones already mentioned) are extensive and suffice to reveal the multi-dimensional significance of the array concept. For instance, although array processing has been extensively used in high frequency communications in the past, the explosive growth in demand for cellular services in recent years has placed it at the centre of interest. Spatial diversity is considered to be one of the most promising solutions for increasing capacity and spectral efficiency. Indeed, the integration of array processing and communications techniques, exploiting the structure of antenna-array systems, has evolved into a well-established technology. This technology is moving from the conventional direction nulling and phase-arrays to advanced *superresolution spatiotemporal-arrays*, MIMO array systems and arrayed wireless sensor networks, which exploit the spatial and temporal properties of the channel in their quest to handle multipaths, and to increase capacity and spectral efficiency. Using these properties, an extra layer of co-channel interference (CCI) and inter-symbol-interference (ISI) cancellation is achieved — asymptotically providing complete interference cancellation.

The performance of array systems, especially the ones with superresolution capabilities is, in general, limited by three main factors:

- The presence of inherent background and sensor noise.
- The limited amount of information the sensors can measure due to finite observation interval (number of snapshots) and array geometry.
- The lack of calibration, modelling errors and system uncertainties that are embedded in the received array signal-vector $\underline{x}(t)$, which are not accounted for. Examples include uncertainties in mutual coupling between sensors, perturbations in the geometrical and electrical characteristics of the array, the presence of moving emitters, nonplanar wavefronts, source angular/temporal spread, etc.

However, the overall quality of the system's performance is naturally a function of the array structure in conjunction with the geometrical characteristics of the signal environment, as well as the algorithms employed. An algorithm would behave differently when used with different array structures and, vice-versa, a certain array would generate different results when its output is applied to different algorithms.

1.1 Nomenclature

It is assumed that the reader is familiar with the fundamentals of vector and matrix algebra. In this book, for typographical convenience, matrices will be denoted by blackboard bold symbols (e.g. $\mathbb{A}, \mathbb{T}, \mathbb{I}$) or, in the absence of a corresponding blackboard bold symbol, by boldface (e.g. $\mathbf{r}$, $\mathbf{k}$, $\mathbf{\Gamma}$). Any underlined symbol will represent a column vector, e.g. $\underline{A}$, $\underline{a}$, $\underline{\mathbf{a}}$. Derivatives with respect to a general parameter p will be denoted with a "dot" (e.g. $\dot{\underline{\mathbf{a}}}$), while the "prime" symbol (e.g. $\underline{\mathbf{a}}'$) will be reserved for differentiation with respect to certain "invariant" parameters. The overall notation to be employed in this book is as follows:

A, a	Scalar
$\underline{A}, \underline{a}$	Column vector
$\mathbb{A}, \mathbf{A}$	Matrix
$(\cdot)^T$	Transpose
$(\cdot)^H$	Hermitian transpose
$(\cdot)^\dagger$	Pseudoinverse
$\|\cdot\|_F$	Frobenius norm of a matrix
$\|\cdot\|$	Norm of a vector
$\lvert\cdot\rvert$	Magnitude
$\odot, \oslash$	Hadamard (Schur) product and division respectively
$\otimes$	Kronecker product
$\exp(\underline{A}$ or $\mathbb{A})$	Elementwise exponential of vector $\underline{A}$ or matrix $\mathbb{A}$
$\mathrm{expm}(\mathbb{A})$	Matrix exponential
$\mathrm{Tr}(\mathbb{A})$	Trace of matrix $\mathbb{A}$
$\det(\mathbb{A})$	Determinant of $\mathbb{A}$
$\mathrm{diag}(\underline{A})$	Diagonal matrix formed from the elements of $\underline{A}$
$\underline{\mathrm{diag}}(\mathbb{A})$	Column vector consisting of the diagonal elements of $\mathbb{A}$
$\mathrm{row}_i\,(\mathbb{A})$	i^{th} row of $\mathbb{A}$
$\mathrm{ele}_{ij}\,(\mathbb{A})$	(i^{th}, j^{th}) element of $\mathbb{A}$
$\mathrm{fix}(A)$	Round down to integer
$\mathcal{E}\{\cdot\}$	Expectation operator
$\underline{A}^b$	Element by element power
$\underline{0}_N$	Zero vector of N elements
$\underline{1}_N$	Column vector of N ones
$\mathbb{I}_N$	$N \times N$ Identity matrix
$\mathbb{O}_{N \times d}$	$N \times d$ Zero matrix
$\mathcal{R}$	Set of real numbers
$\mathcal{N}$	Set of natural numbers
$\mathcal{Z}$	Set of integer numbers
$\mathfrak{C}$	Field of complex numbers

1.2 Main Abbreviations

AGS	Ambiguous Generator Set
ELA	Equivalent Linear Array
CRB	Cramer Rao Bound
DF	Direction Finding
DOA	Directions of Arrival

FOV Field-of-View
SNR Signal-to-Noise Ratio
RMS Root Mean Square
ULA Uniform Linear Array
UCA Uniform Circular Array

1.3 Array of Sensors — Environment

By distributing, in the 3-dimensional Cartesian space, a number $N \geqslant 2$ of sensors (transducing elements, antennas, receivers, etc.) with a common reference point, an array is formed. In general, the positions of the sensors are given by the matrix $\mathbf{r} \in \mathcal{R}^{3 \times N}$

$$\mathbf{r} = \left[\underline{r}_1, \underline{r}_2, \ldots, \underline{r}_N\right] = \left[\underline{r}_x, \underline{r}_y, \underline{r}_z\right]^T \tag{1.1}$$

with $\underline{r}_k = [\mathrm{x}_k, \mathrm{y}_k, \mathrm{z}_k]^T \in \mathcal{R}^{3 \times 1}$ denoting the Cartesian coordinates (location) of the kth sensor of the array $\forall k = 1, 2, \ldots, N$.

It is common practice to express the direction of a wave impinging on the array in terms of the azimuth angle θ, measured anticlockwise from the positive x-axis, and the elevation angle ϕ, measured anticlockwise from the x-y plane, as illustrated in Fig. 1.1. Then, the (3×1) real unit-norm vector

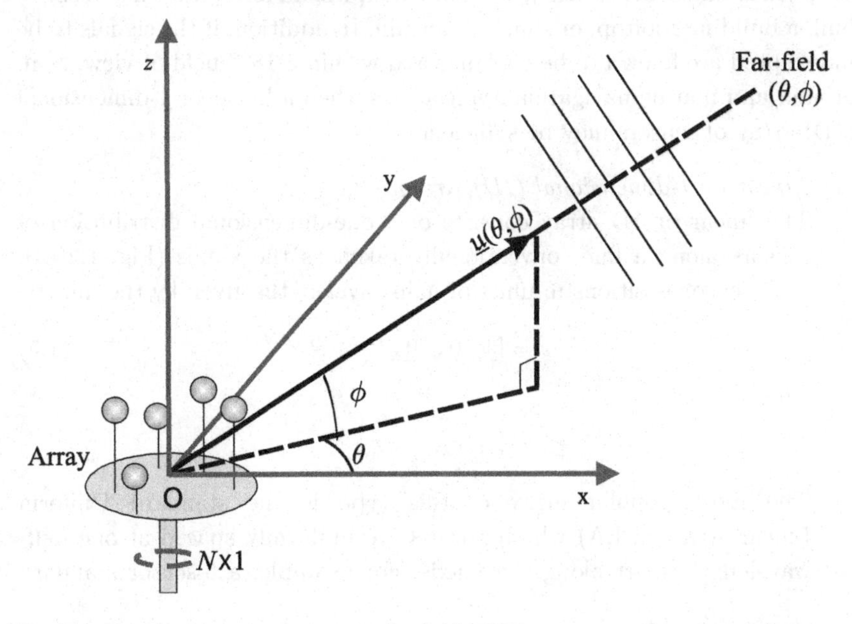

Fig. 1.1 Relative geometry between a far-field emitting source and an array of sensors.

pointing towards the direction (θ, ϕ) is

$$\underline{u} \triangleq \underline{u}(\theta, \phi) = \left[\cos\theta\cos\phi, \sin\theta\cos\phi, \sin\phi\right]^T \tag{1.2}$$

Note that $\|\underline{u}\| = 1$. If the velocity, wavelength and frequency of propagation of the incident wave is denoted by c, λ and F_c, respectively, then the wavenumber vector in the direction (θ, ϕ) is defined as

$$\underline{k} = \underline{k}(\theta, \phi) = \begin{cases} \dfrac{2\pi F_c}{c} \cdot \underline{u} = \dfrac{2\pi}{\lambda} \cdot \underline{u} & \text{in meters} \\ \pi \cdot \underline{u} & \text{in } \lambda/2 \end{cases} \tag{1.3}$$

In the most general case, the parameter space is

$$\Omega = \{(\theta, \phi) : \theta \in [0°, 360°) \quad \text{and} \quad \phi \in (-90°, 90°)\} \tag{1.4}$$

but in most applications, Ω is restricted to only a sector of interest or, in other words, field-of-view (FOV). For instance, in the case of ground surveillance radars, only signals in the plane of the array are of interest — i.e. the system is azimuth-only.

The array configuration is, to a large extent, dictated by the application of interest. One obvious restriction is the shape and size of the available site, which might be, to cite just a few examples, an aircraft's wing, a ship's hull, a building rooftop, or simply a terrain. In addition, if the signals to be intercepted are known to be coplanar and within a 180° field-of-view, as in ground and marine navigation applications, then a linear or 1-dimensional (1D) array of sensors may be sufficient.

(1) *Linear or 1-dimensional (1D) Array.*
The linear or 1D array consists of a one-dimensional distribution of sensors along a line conventionally taken as the x-axis (Fig. 1.2(a)), with sensor positions in units of half-wavelengths given by the matrix

$$\mathbf{r} = [\underline{r}_x, \underline{0}_N, \underline{0}_N]^T \in \mathcal{R}^{3\times N} \tag{1.5}$$

where

$$\underline{r}_x = [r_1, r_2, \ldots, r_N]^T$$

The most popular array of this type is the standard Uniform Linear Array (ULA) whose sensors are uniformly spaced at one half-wavelength apart along the x-axis. For example, a 5-sensor standard

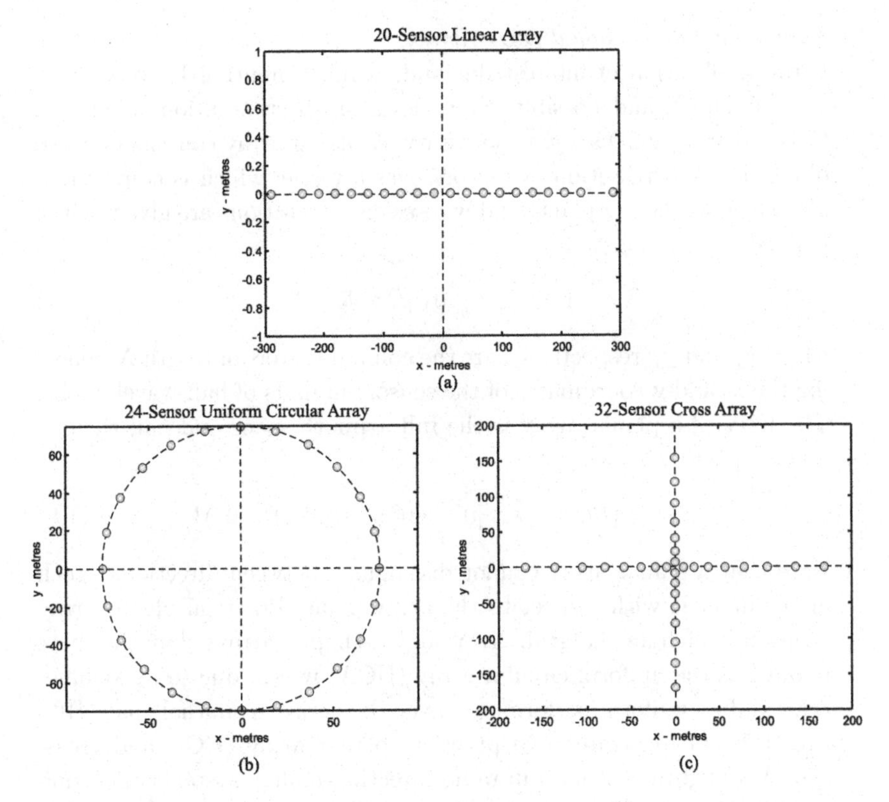

Fig. 1.2 Examples of array geometries: (a) linear array, (b) uniform circular array, (c) "cross" array.

ULA is given by $\underline{r}_x = [-2, -1, 0, 1, 2]^T$ in half wavelengths. Note that the phase reference, or origin of the coordinate system, is taken at the array centroid. By exploiting the regular structure of the ULA, many DF algorithms can be simplified allowing for significant computational savings. The FOV of 1D arrays is restricted to

$$\Omega = \{(\theta, \phi) : \theta \in [0^\circ, 180^\circ), \quad \phi = 0^\circ\} \tag{1.6}$$

It can be easily deduced from the geometry of the problem that a linear array is incapable of distinguishing directions which are symmetric with respect to the array line or which have the same elevation. If a 360° FOV is instead required, then a planar or 2D array should be employed.

(2) *Planar or 2-dimensional (2D) Array.*
If the application of interest demands a full azimuthal FOV, that is, $\theta \in [0°, 360°)$, and possibly some elevation discrimination capability, then a planar or 2D array is necessary. A planar array consists of a two dimensional distribution of sensors over a plane, which is conventionally taken as the x-y plane, and whose sensor positions are given by the matrix

$$\mathbf{r} = [\underline{r}_x, \underline{r}_y, \underline{0}_N]^T \in \mathcal{R}^{3\times N} \tag{1.7}$$

where $\underline{r}_x$ and $\underline{r}_y$, respectively, are the column vectors of length N denoting the x- and y-coordinates of the sensors in units of half-wavelengths. The FOV of a planar array is the full azimuth space and half the elevation space

$$\Omega = \{(\theta, \phi) : \theta \in [0°, 360°), \quad \phi \in [0°, 90°)\} \tag{1.8}$$

Note that a planar array cannot distinguish between directions which are symmetric with respect to the array plane. Practical planar array structures include the grid, X, Y and L-shaped arrays, but the most popular is the uniform circular array (UCA) which, due to its symmetry, exhibits uniform performance over the entire azimuthal space [16]. Fig. 1.2(b,c) illustrates two practical planar array (UCA and cross-array) configurations used in radio direction-finding systems operating in the UK in the HF (3 to 30 MHz) band. It is important to point out that a planar array also permits the discrimination of signals in elevation $\phi \in [0°, 90°)$ — i.e. the emitters need not be coplanar (as is the case in airborne surveillance), although its resolving power in the elevation space is not as good as that of a 3D array. A 3D array has a FOV spanning the entire parameter space. However, for most applications 1D and 2D arrays prove to be sufficient.

1.4 Pictorial Notation

1.4.1 *Spaces/Subspaces*

Because the visualization of a space greater than a 3-dimensional real space is impossible, the notation shown in Fig. 1.3 is employed to provide some illustrative aid. However, this pictorial representation should be used with care.

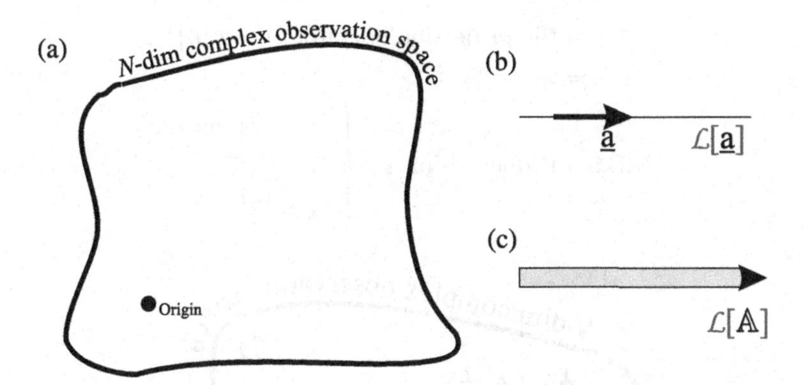

Fig. 1.3 Illustrative Notation: (a) denotes a N-dimensional complex (or real) observation space. Note that any vector in this space has N elements; (b) denotes a one-dimensional subspace/space spanned by the vector $\underline{\mathbf{a}}$; (c) denotes an M-dimensional subspace/space (with $M \geq 2$) spanned by the columns of the matrix $\mathbb{A}$.

1.4.2 *Projection Operator*

Consider an $(N \times M)$ matrix $\mathbb{A}$ with $M < N$ (i.e. the matrix has M columns) and let $\mathcal{L}[\mathbb{A}]$ represent the linear subspace spanned by the columns of $\mathbb{A}$. Assuming that the columns of $\mathbb{A}$ are linearly independent — that is, a column of $\mathbb{A}$ cannot be written as a linear combination of the remaining $M-1$ columns — the subspace $\mathcal{L}[\mathbb{A}]$ is of dimensionality M, i.e.

$$\dim \{\mathcal{L}[\mathbb{A}]\} = M < N \tag{1.9}$$

lying in a N-dimensional space $\mathcal{H}$, as shown in Fig. 1.4. The complement subspace to $\mathcal{L}[\mathbb{A}]$ is denoted by $\mathcal{L}[\mathbb{A}]^{\perp}$ and is of dimensionality $N - M$, i.e.

$$\dim \left\{\mathcal{L}[\mathbb{A}]^{\perp}\right\} = N - M \tag{1.10}$$

Then:

(1) any vector $\underline{x} \in \mathcal{L}[\mathbb{A}]$ can be written as a linear combination of the columns of the matrix $\mathbb{A}$
(2) any vector $\underline{x} \in \mathcal{H}$ can be projected onto $\mathcal{L}[\mathbb{A}]$ (or onto $\mathcal{L}[\mathbb{A}]^{\perp}$), as shown in Fig. 1.5, using the concept of the projection operator defined as follows:

$$\begin{aligned}\mathbb{P}_{\mathbb{A}} &= \text{projection operator onto the subspace } \mathcal{L}[\mathbb{A}] \\ &= \mathbb{A}(\mathbb{A}^H \mathbb{A})^{-1}\mathbb{A}^H \end{aligned} \tag{1.11}$$

$$\begin{aligned}\mathbb{P}_{\mathbb{A}}^{\perp} &= \text{the projection operator onto } \mathcal{L}[\mathbb{A}]^{\perp} \\ &= \mathbb{I}_N - \mathbb{P}_{\mathbb{A}} \end{aligned} \tag{1.12}$$

$$\text{N.B.:} \quad \text{Properties of } \mathbb{P}_{\mathbb{A}} \begin{cases} (N \times N) \text{ matrix} \\ \mathbb{P}_{\mathbb{A}} \mathbb{P}_{\mathbb{A}} = \mathbb{P}_{\mathbb{A}} \\ \mathbb{P}_{\mathbb{A}} = \mathbb{P}_{\mathbb{A}}^H \end{cases} \tag{1.13}$$

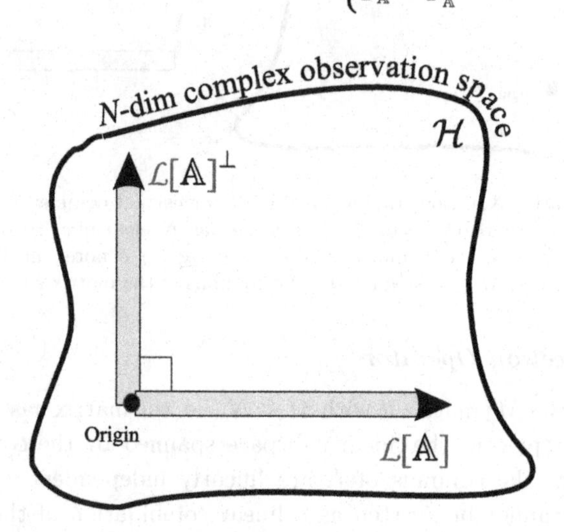

Fig. 1.4 Illustrative representation of the subspaces $\mathcal{L}[\mathbb{A}]$ and $\mathcal{L}[\mathbb{A}]^{\perp}$ lying in an N-dimensional complex space.

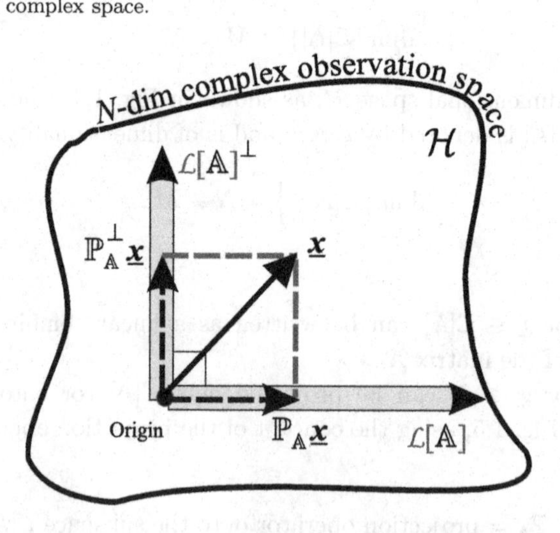

Fig. 1.5 Illustrative representation of the projection of any vector $\underline{x} \in \mathcal{H}$ onto $\mathcal{L}[\mathbb{A}]$ and onto $\mathcal{L}[\mathbb{A}]^{\perp}$.

(3) For any vector $\underline{x} \in \mathcal{L}[\mathbb{A}]$ the following expressions are valid

$$\begin{cases} \mathbb{P}_{\mathbb{A}}\,\underline{x} = \underline{x} \\ \mathbb{P}_{\mathbb{A}}^{\perp}\,\underline{x} = \underline{0}_N \end{cases} \tag{1.14}$$

1.5 Principal Symbols

In this book the following symbols will always denote

p, q	generic parameters
θ	azimuth angle
ϕ	elevation angle
α	alpha — cone angle
β	beta — cone angle
s	arc length
N	number of sensors
$\mathbf{r}$	$(3 \times N)$ real matrix with columns the sensor locations
$\underline{\mathbf{a}}$	array manifold vector
l_{m}	array manifold length
l_{a}	array aperture
$\mathcal{A}$	array manifold curve
$\mathcal{M}$	array manifold surface
$\mathcal{H}$	space/subspace

1.6 Modelling the Array Signal-Vector and Array Manifold

Consider an array of N sensors, with sensor locations $\mathbf{r}$, operating in the presence of M narrowband point sources and having the same known carrier frequency F_c. The modelling of the signal due to the ith emitter, received at the zero-phase reference point (taken to be the origin of the coordinate system) is determined by whether the source is located in, or close to, the array's *near-field* or in the array's *far-field.* In practice, this is determined according to the value of its range ρ_i with respect to the array aperture l_{a}, defined as follows:

$$l_{\mathrm{a}} = \max_{\forall i,j} \left\| \underline{r}_i - \underline{r}_j \right\| \tag{1.15}$$

To be more specific, if $\rho_i \simeq 2l_a/\lambda$, where λ is the wavelength of the transmitted signals, the ith source is located close to the array near-field border (which defines the so-called *Fresnel* zone) and the spherical wave propagation model has to be considered. For $\rho_i > 2l_a/\lambda$ and especially for $\rho_i \gg 2l_a/\lambda$, the ith source is situated in the array far-field (the so-called *Fraunhofer* zone) and plane wave propagation is assumed. As a matter of fact, sources of range $\rho_i \simeq 2l_a/\lambda$ or $\rho_i > 2l_a/\lambda$, but not $\rho_i \gg 2l_a/\lambda$, are usually regarded as being located in the *near far-field* of an array, for which the spherical wave propagation model has to be utilized.

Based on the above discussion and by considering the M sources in the far-field of the array, the array signal is the superposition of plane waves from each individual source. The planewave/signal due to the ith emitter, received at the zero-phase reference point, can be written as

$$i\text{th signal at the reference point: } m_i(t)\exp(j2\pi F_c t) \tag{1.16}$$

where $m_i(t)$ represents the complex envelope (message) of the signal and $\exp(j2\pi F_c t)$ represents the carrier. If τ_{ik} represents the propagation delay of the ith signal between the phase reference location and the kth sensor then the signal arriving at the kth sensor will be a delayed version of the signal given in Eq. (1.16), expressed as shown below

$$i\text{th signal at the } k\text{th sensor: } \underbrace{m_i(t-\tau_{ik})}_{\approx m_i(t)}\exp(j2\pi F_c(t-\tau_{ik})) \tag{1.17}$$

This propagation delay τ_{ik} is a function of the DOA of the ith signal and the position of the kth sensor with respect to the reference point. Indeed, with reference to Fig. 1.6, τ_{ik} can be derived as

$$\begin{aligned}
\tau_{ik} &= \frac{\ell_{ik}}{c} = \frac{\left\|\mathbb{P}_{\underline{u}_i}\,\underline{r}_k\right\|}{c} \\
&= \frac{\sqrt{\underline{r}_k^T\,\mathbb{P}_{\underline{u}_i}^T\,\mathbb{P}_{\underline{u}_i}\,\underline{r}_k}}{c} \\
&= \frac{\sqrt{\underline{r}_k^T\,\mathbb{P}_{\underline{u}_i}\,\underline{r}_k}}{c} \quad \text{(using Eq. (1.13))} \\
&= \frac{\sqrt{\underline{r}_k^T\,\underline{u}_i\left(\underline{u}_i^T\,\underline{u}_i\right)^{-1}\underline{u}_i^T\,\underline{r}_k}}{c}
\end{aligned}$$

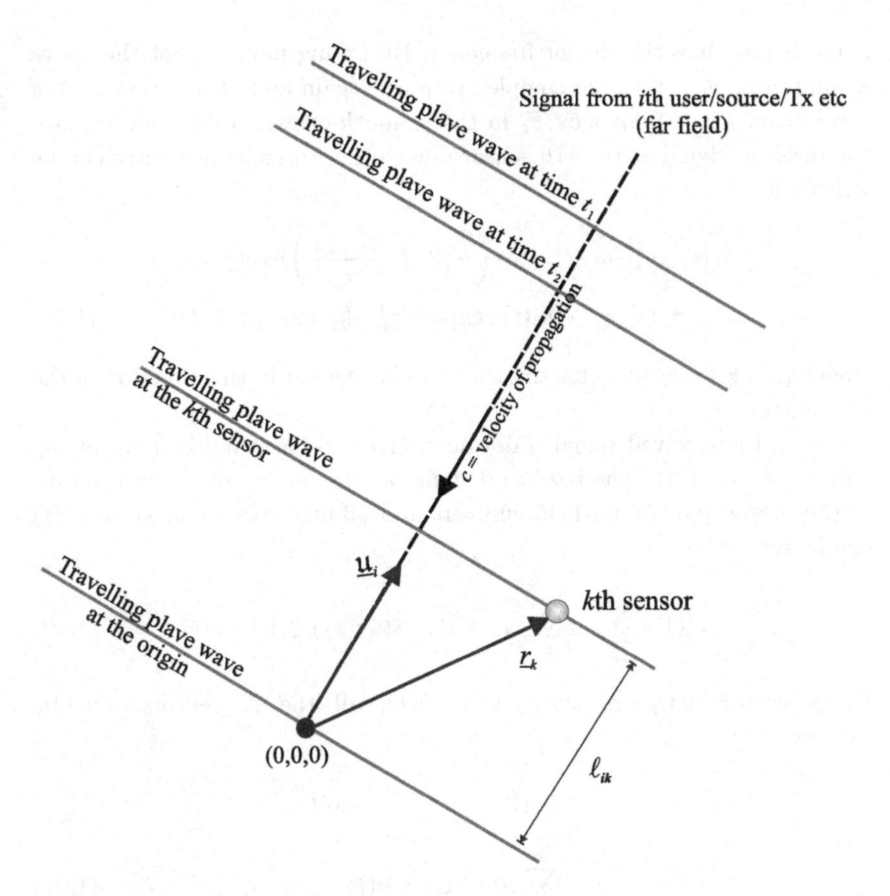

Fig. 1.6 Geometry of a travelling plane wave, relative to the kth sensor and the array reference point.

$$= \frac{\sqrt{\underline{r}_k^T \underline{u}_i \underline{u}_i^T \underline{r}_k}}{c} = \frac{\sqrt{\left(\underline{r}_k^T \underline{u}_i\right)^2}}{c} = \frac{\underline{r}_k^T \underline{u}_i}{c}$$

i.e.

$$\tau_{ik} = \frac{\underline{r}_k^T \underline{u}_i}{c} \tag{1.18}$$

It is reasonable to assume that the envelope (baseband signal $m_i(t)$) does not change significantly as it traverses the array. Hence $\tau_{ik} \leqslant \tau_{\max}, \forall i, k$ and $m_i(t - \tau_{\max}) \simeq m_i(t)$ has been used in Eq. (1.17), where $\tau_{\max}$ is the maximum possible time for a signal to traverse the array. This is due to the narrowband assumption, i.e. the highest frequency in the message signal is much less than the carrier frequency. By taking into account the above assumption, as well as the complex response (gain and phase) $\gamma_k(\theta_i, \phi_i)$ of the kth sensor at frequency F_c in the azimuth-elevation direction (θ_i, ϕ_i), the received signal at the kth sensor due to the ith emitting source can be written as

$$\begin{aligned}\gamma_k(\theta_i, \phi_i) \cdot m_i(t) \cdot \exp\left(-j2\pi F_c \frac{\underline{r}_k^T \cdot \underline{u}_i}{c}\right) \exp(j2\pi F_c t) \\ = \gamma_k(\theta_i, \phi_i) \cdot m_i(t) \cdot \exp(-j\underline{r}_k^T \cdot \underline{k}_i) \exp(j2\pi F_c t)\end{aligned} \tag{1.19}$$

where $\underline{k}_i = \underline{k}(\theta_i, \phi_i)$ denotes the wavenumber vector in the direction of the ith emitter.

Next, the received signal is downconverted to baseband by multiplying with $\exp(-j2\pi F_c t)$. The baseband signal at the output of the kth sensor, in the presence of M far-field emitters and additive baseband noise $n_k(t)$ can be written as

$$x_k(t) = \sum_{i=1}^{M} \gamma_k(\theta_i, \phi_i)\, m_i(t) \exp(-j\, \underline{r}_k^T \underline{k}_i) + n_k(t) \tag{1.20}$$

Using vector notation, the output from all the N sensors, can be expressed as

$$\begin{aligned}\underline{x}(t) &= \left[x_1(t), x_1(t), \ldots, x_N(t)\right]^T \\ &= \sum_{i=1}^{M} m_i(t)\, \underline{\mathbf{a}}_i + \underline{n}(t)\end{aligned} \tag{1.21}$$

where $\underline{n}(t) = \left[n_1(t)\ n_2(t)\ \cdots\ n_N(t)\right]^T \in \mathfrak{C}^{N\times 1}$ denotes the baseband additive white Gaussian noise of power σ^2. The $(N \times 1)$ complex vector $\underline{\mathbf{a}}_i$ is known as the array manifold vector associated with the ith emitting source (signal), representing the complex array response to a unit amplitude plane wave impinging from direction (θ_i, ϕ_i), compactly written as

$$\underline{\mathbf{a}}_i \triangleq \underline{\mathbf{a}}(\theta_i, \phi_i) = \underline{\gamma}(\theta_i, \phi_i) \odot \exp\left(-j \underbrace{[\underline{r}_x, \underline{r}_y, \underline{r}_z]}_{=\mathbf{r}^T} \underline{k}(\theta_i, \phi_i)\right) \tag{1.22}$$

where

$$\underline{\gamma}(\theta_i, \phi_i) = [\gamma_1(\theta_i, \phi_i), \gamma_2(\theta_i, \phi_i), \ldots, \gamma_N(\theta_i, \phi_i)]^T \in \mathfrak{C}^{N\times 1}$$

This vector is also referred to as the "array response vector" or "array steering vector."

Given that the same array may be represented by an infinite number of matrices **r** through a change in the coordinate system reference point, a choice is made to fix the coordinate reference point $(0, 0, 0)$ to be the array centroid. This translates into a condition on the location vectors $\underline{r}_x$, $\underline{r}_y$ and $\underline{r}_z$ such that

$$\text{sum}(\underline{r}_x) = \text{sum}(\underline{r}_y) = \text{sum}(\underline{r}_z) = 0 \tag{1.23}$$

or equivalently,

$$\underline{r}_x^T \underline{1}_N = \underline{r}_y^T \underline{1}_N = \underline{r}_z^T \underline{1}_N = 0 \tag{1.24}$$

As can be seen from Eq. (1.22), a shift in the reference point of the array coordinate system results in a manifold vector $\underline{\mathbf{a}}_r(\theta, \phi)$ which is related to the initial manifold vector:

$$\underline{\mathbf{a}}_r(\theta, \phi) = g(\theta, \phi) \cdot \underline{\mathbf{a}}(\theta, \phi) \tag{1.25}$$

where $g(\theta, \phi)$ is a complex number with $|g(\theta, \phi)| = 1, \forall(\theta, \phi)$. In fact, $g(\theta, \phi) = \exp(-j\underline{r}_0^T \underline{k}(\theta, \phi))$ where $\underline{r}_0$ is the translation vector of the reference point from the centroid of the array. For simplicity, all arrays henceforth will have their centroid as the coordinate reference point, i.e. $g(\theta, \phi) = 1, \forall(\theta, \phi)$.

Note that in the case of a linear array of isotropic sensors Eq. (1.22) simplifies to

$$\underline{\mathbf{a}}(\theta) = \exp(-j\pi\underline{r}_x \cos\theta) \tag{1.26}$$

while for a planar array it takes the following form:

$$\underline{\mathbf{a}}(\theta, \phi) = \exp(-j\pi\underline{r}(\theta)\cos\phi) \tag{1.27}$$

where

$$\underline{r}(\theta) = \underline{r}_x \cos\theta + \underline{r}_y \sin\theta \tag{1.28}$$

It can easily be seen that $\underline{r}(\theta)$ in Eq. (1.27) represents the positions of the projections of the sensor locations of the planar array onto the line of azimuth θ (see Fig. 1.7), and is known as the "Equivalent Linear Array" (ELA) along the direction θ. Notice the similarity between the response

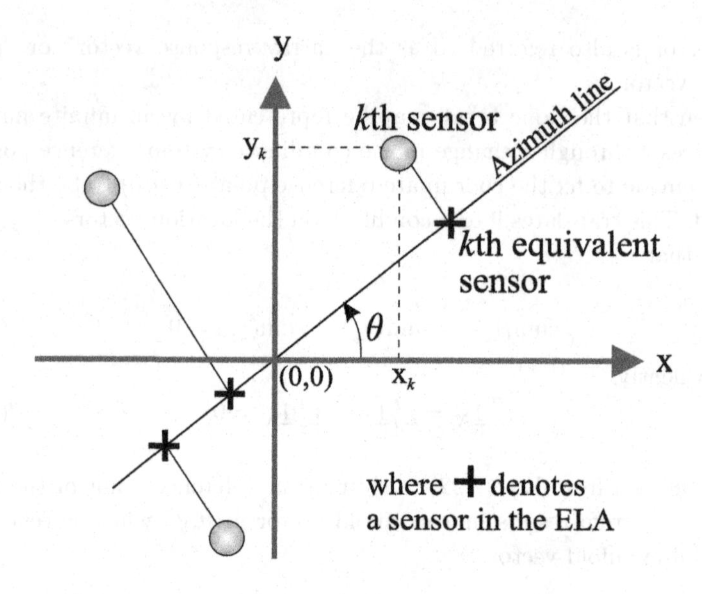

Fig. 1.7 Equivalent Linear Array (ELA) associated with the azimuth angle θ.

vectors of the linear and planar arrays in Eqs. (1.26) and (1.27) respectively. Using Eq. (1.22) the observed array signal-vector $\underline{x}(t) \in \mathfrak{C}^{N \times 1}$ of Eq. (1.21) can be written concisely as

$$\underline{x}(t) = \mathbb{A}\underline{m}(t) + \underline{n}(t) \tag{1.29}$$

where

$$\begin{cases} \mathbb{A} = [\underline{\mathbf{a}}_1, \underline{\mathbf{a}}_2, \ldots, \underline{\mathbf{a}}_M] \\ \underline{m}(t) = [m_1, m_2, \ldots, m_M]^T \end{cases} \tag{1.30}$$

where $\mathbb{A} \in \mathfrak{C}^{N \times M}$ is the array manifold (response) matrix and $\underline{m}(t) \in \mathfrak{C}^{M \times 1}$ is the vector of signal envelopes (messages). Note that in the case of isotropic sensors the matrix $\mathbb{A}$ can be written in a compact way as

$$\mathbb{A} = \exp(-j\mathbf{r}^T\mathbf{k}) \tag{1.31}$$

with $\mathbf{k} = [\underline{\mathbf{k}}_1, \underline{\mathbf{k}}_2, \ldots, \underline{\mathbf{k}}_M]$ representing the wavenumber matrix.

Based on the above model of Eq. (1.29), the theoretical covariance matrix $\mathbb{R}_{xx}$ of the array signal-vector $\underline{x}(t)$ can be formed as

$$\begin{aligned} \mathbb{R}_{xx} &\triangleq \mathcal{E}\{\underline{x}(t)\,\underline{x}(t)^H\} \in \mathfrak{C}^{N \times N} \\ &= \mathbb{A}\mathbb{R}_{mm}\mathbb{A}^H + \sigma^2\mathbb{I}_N \end{aligned} \tag{1.32}$$

where $\mathbb{R}_{mm} \triangleq \mathcal{E}\{\underline{m}(t)\,\underline{m}(t)^H\} \in \mathfrak{C}^{M\times M}$ is the source covariance matrix, which is diagonal in the case of uncorrelated signals and $\sigma^2\mathbb{I}_N$ is the additive white Gaussian noise covariance matrix $\mathbb{R}_{nn} \triangleq \mathcal{E}\{\underline{n}(t)\,\underline{n}(t)^H\} \in \mathfrak{C}^{N\times N}$. In practice, only a finite number L data snapshots $[\underline{x}(t_1), \underline{x}(t_2), \ldots, \underline{x}(t_L)]$, collected by an array of sensors, can be observed. Hence only an estimate of $\mathbb{R}_{xx}$ is computed in the form of the sample covariance matrix given by

$$\widehat{\mathbb{R}}_{xx} \triangleq \frac{1}{L}\sum_{i=1}^{L} \underline{x}(t_i)\,\underline{x}(t_i)^H = \frac{1}{L}\mathbb{X}\mathbb{X}^H \tag{1.33}$$

where

$$\mathbb{X} \triangleq [\underline{x}(t_1), \underline{x}(t_2), \ldots, \underline{x}(t_L)] \in \mathfrak{C}^{N\times L}$$

It is important to point out that if the assumption $m_i(t-\tau_{ik}) \approx m_i(t), \forall i, k$ used in Eq. (1.17) is *not* valid, then the array signal vector associated with the ith signal should be modelled as

$$\underline{x}(t) = \sum_{i=1}^{M} \underline{m}_{\tau_i}(t) \odot \underline{\mathbf{a}}_i + \underline{n}(t) \tag{1.34}$$

where

$$\underline{m}_{\tau_i}(t) = [m_i(t-\tau_{i1}), m_i(t-\tau_{i2}), \ldots, m_i(t-\tau_{iN})]^T$$

1.7 Significance of Array Manifolds

It is clear from Eqs. (1.22), (1.26) or (1.27) that the manifold vector for a particular direction contains all the information about the geometry involved when a wave is incident on the array from that direction. By recording the locus of the manifold vectors as a function of direction, a "continuum" (i.e. a geometrical object such as a curve or surface) is formed lying in an N-dimensional space. This geometrical object (locus of manifold, or response, vectors) is known as the *array manifold.* The array manifold can be calculated (and stored) from only the knowledge of the locations and directional characteristics of the sensors. Thus, according to Schmidt [1], "the array manifold completely characterizes any array and provides a representation of the real array into N-dimensional complex space."

The significance of the array manifold concept becomes apparent when it is realized that all subspace-based parameter estimation algorithms involve searching over the array manifold for response vectors which satisfy a given criterion. For example, in the case of the MUSIC algorithm [7], the manifold is searched for the array manifold vectors which are (nearly) orthogonal to the estimated noise subspace. In other words the manifold is searched for vectors belonging to the subspace $\mathcal{L}[\mathbb{A}]$ i.e. the subspace spanned by the columns of the matrix $\mathbb{A} = [\underline{\mathbf{a}}_1, \underline{\mathbf{a}}_2, \ldots, \underline{\mathbf{a}}_M]$ (the signal subspace). Figure 1.8 illustrates a single-parameter manifold curve embedded in N-dimensional complex space together with the subspace $\mathcal{L}[\mathbb{A}]$ associated with two sources. These two sources can be asymptotically resolved, irrespective of their angular separation — hence the title "superresolution." It is clear that the accuracy of the DOA estimates is dependent on how accurately the noise subspace is estimated; which in turn is a function of the observation interval and the signal-to-noise ratio (SNR). Thus the DOAs can be exactly estimated from the array output only on the basis of either an infinite number of snapshots (i.e. $L \to \infty$) or, an infinite signal-to-noise ratio scenario (i.e. $n(t) = 0$) — asymptotic conditions.

Another determining factor which might not be initially apparent is the shape of the array manifold. A little thought reveals that a potentially unresolvable situation might arise if two response (manifold) vectors corresponding to different DOAs happen to be identical. For instance, this "abnormality" arises when a single-parameter array manifold crosses upon itself, as shown in Fig. 1.9. Such phenomena are commonly referred to as "ambiguities" and are a direct consequence of the array geometry dictating

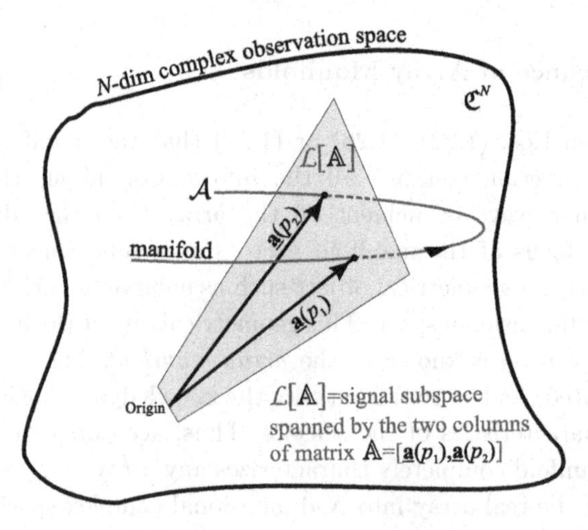

Fig. 1.8 Illustrative representation of an array manifold curve embedded in N-dimensional complex space and the subspace spanned by two manifold vectors.

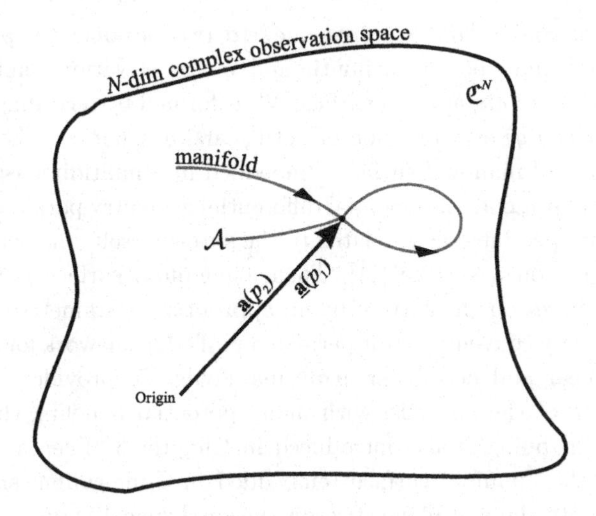

Fig. 1.9 Illustrative representation of array "abnormality" (ambiguity).

the behavior of the array manifold. As will be discussed below, the behavior of the array manifold, in particular its local geometry, plays an important role in handling "abnormalities" and also in defining the capabilities (e.g. the resolving power) of an array system.

1.8 An Outline of the Book

The theoretical framework associated with curves lying in N-dimensional complex space, qualifying to be manifolds of linear array structures, is presented in Chapter 2. More specifically by recording the locus of the vector $\underline{a}(p)$ as a function of p, a one-dimensional *continuum* (i.e. a curve $\mathcal{A}$) is formed known as the array manifold and embedded in a N-dimensional space. Thus in this chapter the properties and characteristics of single-parameter manifolds (i.e. curves) are investigated and supported by a number of representative examples of symmetric and non-symmetric linear array geometries [10]. These manifold curves have been found to have a hyper-helical shape with numerous advantages. For example, all the curvatures of a hyperhelix are constant (do not vary from point to point) and may be evaluated recursively. The convenient nature of a hyperhelix's geometry will be proven invaluable for the rest of the book, not only for linear but also for non-linear (2D and 3D) arrays.

If the manifold vector $\underline{\mathbf{a}}$ corresponds to two parameters (p,q) where, for instance p may be the azimuth and q the elevation angle, then a 2-dimensional *continuum*, i.e. a surface $\mathcal{M}$, is formed by recording the locus of the vector $\underline{\mathbf{a}}(p,q)$ as a function of both p and q. Chapter 3 is concerned with the study of manifold surfaces embedded in a multidimensional complex space and presents the essential differential geometry parameters which have been grouped into those related to the surface itself, and those related to curves lying on a surface [11]. Then a manifold surface is treated as a family of curves on the surface by an appropriate parametrization. This treatment is very convenient as it permits a unified framework for the analysis of the linear and non-linear array manifolds. To provide a simplified representation of the analysis, with many potential benefits, the concept of isometric mapping is also introduced in Chapter 3. Then an isometric mapping of the manifold surface (embedded in a multidimensional complex space) onto the real plane (two-dimensional space) is presented which preserves certain differential geometry characteristics of the manifold surface, under certain conditions.

It is common practice, and intuitively appealing, in array processing to use azimuth and elevation as the directional parameters of a waveform impinging on an array of sensors. However, this is by no means unique and, furthermore, this is not the most suitable parametrization for the study of the behavior of the array manifold [12]. Therefore in Chapters 4 and 5, based on the material presented in Chapters 2 and 3, the following manifold-surface parametrizations

- the (azimuth, elevation), or (θ, ϕ), parametrization
- the cone-angles, or (α, β), parametrization

are examined for non-linear arrays (2D and 3D geometries). The significance of these parametrizations is demonstrated by a number of representative examples. Furthermore, properties such as Gaussian and geodesic curvatures are defined and their implications with regards to isometric mappings are discussed.

In the next two chapters of the book (Chapters 6 and 7) the fundamental effects of the array geometry behavior on the performance of the system irrespective of the type of algorithm used, are studied. The array structure is incorporated into the observed array signal-vectors $\underline{x}(t)$ (and therefore to the estimation problem) through the array manifold. Thus the geometry of an array plays a crucial role in dictating the shape, properties and "anomalies" of the array manifold, and as a consequence, in dictating the phenomenon where some manifold vectors can be written as linear combinations of others [13,25]. This gives rise to the array "ambiguity problem" where the occurrence of *false* parameter estimates seriously impairs the performance of array system irrespective of its other capabilities. In particular, in Chapters 6 and 7 two general types of ambiguities are presented based on the partitioning of the array manifold curves into *equal* or *unequal* segments from which the ambiguous generator sets (AGS) can be constructed — with each AGS representing an infinite number of ambiguous sets of parameter values (e.g. directions). Furthermore, the theoretical aspects of the investigation provide a sufficient condition for the presence of ambiguities while the results are then extended to non-linear arrays by treating surfaces as families of curves [14].

Finally, in Chapter 8, the knowledge of the shape of the array manifold is used to determine the array's ultimate capabilities according to the following criteria:

- Accuracy and the Cramer Rao Lower Bound
- Detection threshold
- Resolution threshold

In particular, by approximating, locally, a manifold curve with a circular arc, the Cramer Rao Lower Bound (CRB) is studied and its relation to resolution and detection bounds is established [15, 16]. Then by defining the detection and resolution subspaces, in conjunction with the circular approximation (locally) of the array manifold, the minimum arc length separation in order to detect and resolve two sources located close together is estimated. This is done in terms of the curve's principal curvature only, thereby simplifying the analysis considerably.

Chapter 2

Differential Geometry of Array Manifold Curves

2.1 Manifold Curve Representation — Basic Concepts

Let $\underline{\mathbf{a}} \triangleq \underline{\mathbf{a}}(p) \in \mathfrak{C}^N$ be the manifold vector of an array of N sensors where p is a generic directional parameter. This is a single-parameter *vector function* and as p varies the point $\underline{\mathbf{a}}$ will trace out a curve $\mathcal{A}$, as shown in Fig. 2.1, embedded in an N-dimensional complex space $\mathfrak{C}^N$. This was expected, as vector functions of one parameter are used to define space curves — also known as single-parameter manifolds.

The curve $\mathcal{A}$, which is formally defined

$$\mathcal{A} \triangleq \{\underline{\mathbf{a}}(p) \in \mathfrak{C}^N, \ \forall p : p \in \Omega_p\} \tag{2.1}$$

where Ω_p denotes the parameter space, is said to be a *regular parametrized differential curve* if

$$\dot{\underline{\mathbf{a}}}(p) \neq \underline{0}_N, \quad \forall p \in \Omega_p \tag{2.2}$$

where a "dot" at the top of a symbol is used to denote differentiation with respect to parameter p. As the vector $\dot{\underline{\mathbf{a}}}(p)$ represents the tangent vector to the curve, the "regularity" condition of Eq. (2.2) ensures that the tangent vector exists at all points on the array manifold. The arc length $s(p)$ along the manifold curve $\mathcal{A}$ and its rate-of-change $\dot{s}(p)$ are formally defined, respectively, as

$$s(p) \triangleq \int_0^p \|\dot{\underline{\mathbf{a}}}(p)\| \, dp \tag{2.3}$$

and

$$\dot{s}(p) = \frac{ds}{dp} \triangleq \|\dot{\underline{\mathbf{a}}}(p)\| \tag{2.4}$$

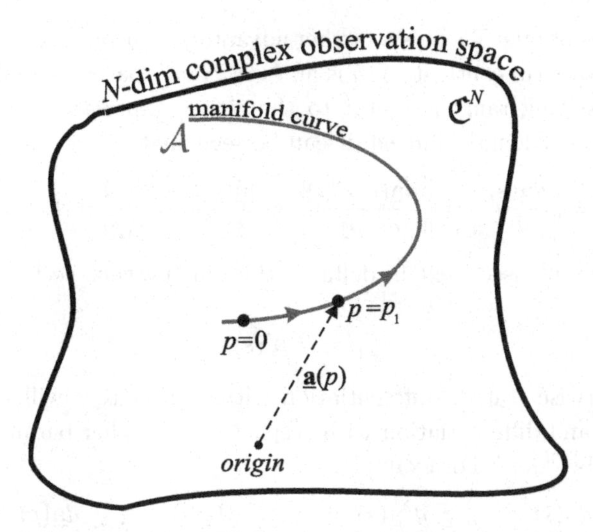

Fig. 2.1 Manifold curve $\mathcal{A}$ embedded in $\mathfrak{C}^N$.

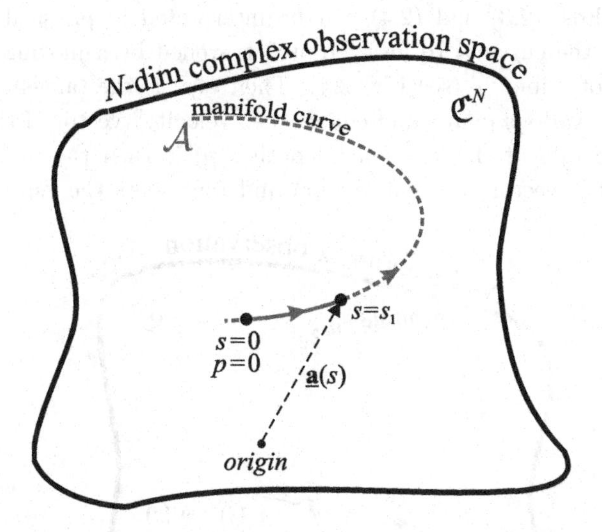

Fig. 2.2 Manifold parameterization in terms of arc length s.

The array manifold is conventionally parametrized in terms of a generic bearing parameter p. For instance for a linear array, p may represent the azimuth angle θ. However, parametrization in terms of the arc length s (see Fig. 2.2), which is the most basic feature of a curve and a natural parameter representing the actual physical length of a segment of the manifold curve in

$\mathfrak{C}^N$, is more suitable. There is a further advantage of using s as a parameter: the arc length s (in contrast to p) is an *"invariant"* parameter. This means that the resulting tangent vector to the curve, expressed in terms of s, always has unity length. Indeed, it can be seen that

$$\|\underline{\mathbf{a}}'(s)\| = \left\|\frac{d\underline{\mathbf{a}}(s)}{ds}\right\| = \left\|\frac{d\underline{\mathbf{a}}(p)/dp}{ds/dp}\right\| = \frac{\|\dot{\underline{\mathbf{a}}}(p)\|}{\dot{s}(p)} = \frac{\dot{s}(p)}{\dot{s}(p)} = 1, \quad \forall s \tag{2.5}$$

which is a result used next in defining the unit tangent vector $\underline{u}_1(s)$ (see Fig. 2.3)

$$\underline{u}_1(s) \triangleq \underline{\mathbf{a}}'(s) \tag{2.6}$$

Unless otherwise stated, differentiation with respect to s will be denoted by "prime" and differentiation with respect to any other parameter p will be denoted by "dot." For example,

$$\underline{\mathbf{a}}' \triangleq \frac{d\underline{\mathbf{a}}(s)}{ds}; \quad \underline{\mathbf{a}}'' \triangleq \frac{d^2\underline{\mathbf{a}}(s)}{ds^2}; \quad \underline{\mathbf{a}}''' \triangleq \frac{d^3\underline{\mathbf{a}}(s)}{ds^3}; \quad \dot{\underline{\mathbf{a}}} \triangleq \frac{d\underline{\mathbf{a}}(p)}{dp} \tag{2.7}$$

Note that Eqs. (2.3) and (2.4) can be interpreted in physical terms by considering the curve $\mathcal{A}$ to be a "route" travelled by a moving object as a function of "time" p (see Fig. 2.4). Then $\dot{\underline{\mathbf{a}}}(p)$ is the tangent vector to the curve at various points and equals the "velocity" vector of the moving object. The rate of change of arc length $\dot{s}(p)$ is then the magnitude of the "velocity" vector (tangent vector) and represents the "speed" of the

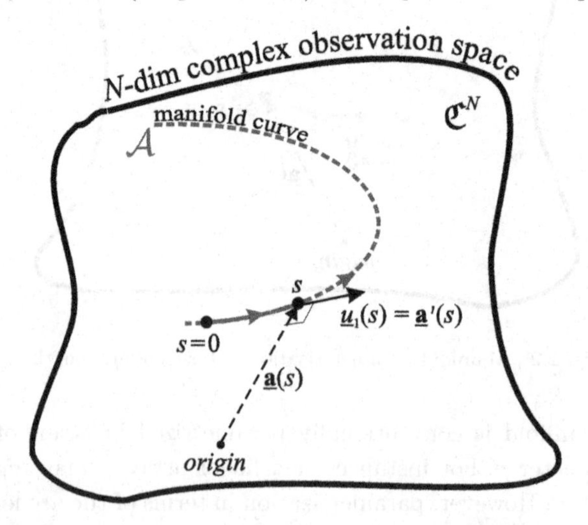

Fig. 2.3 Arc length s is an "invariant" parameter — tangent vector $\underline{u}_1(s)$ has unit norm $\forall s$.

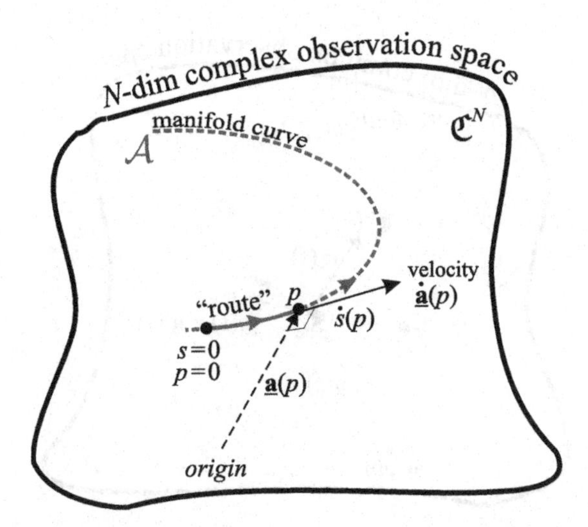

Fig. 2.4 The parameter p is not an invariant parameter — i.e. $speed = \|\dot{\underline{\mathbf{a}}}(p)\| = \dot{s}(p) = variable \neq 1$.

moving object while the total distance travelled, $s(p)$, can be obtained by integrating speed with respect to "time" p. Parametrization in terms of s, i.e. $\underline{\mathbf{a}}(s)$, indicates that the "velocity" vector (tangent vector) $\underline{\mathbf{a}}'(s)$ always has constant length equal to 1 (moving object with unity speed).

Note that, in general, it is not necessary to mention the origin of the arc length ($s = 0$) since most concepts are defined only in terms of the derivatives of $\underline{\mathbf{a}}(s)$. Furthermore, the rate of change of arc length $\dot{s}(p)$ is a local property of the curve and, as will be demonstrated in Chapter 8, plays a crucial role in dictating the resolution/detection capabilities offered by the array manifold and consequently by the geometry of the array.

As was mentioned before, the tangent vector $\underline{\mathbf{a}}'(s)$ is of unity length $\forall s$, indicating that s is an "invariant" parameter. In this case the norm of the second derivative of the manifold vector (parametrized in terms of s) $\|\underline{\mathbf{a}}''(s)\|$ measures how fast the curve pulls away from the tangent line at s, in the neighborhood of s (see Fig. 2.5). In other words, it measures the rate of change of the angle which neighboring tangents make with the tangent at s. The number $\|\underline{\mathbf{a}}''(s)\| = \kappa_1(s)$ is called the first curvature of the manifold curve at s. As we will see next, although many curvatures can be defined at a point on the manifold curve $\mathcal{A} \in \mathfrak{C}^N$, the first curvature is the most important one and for this reason this is also known as "principal" curvature.

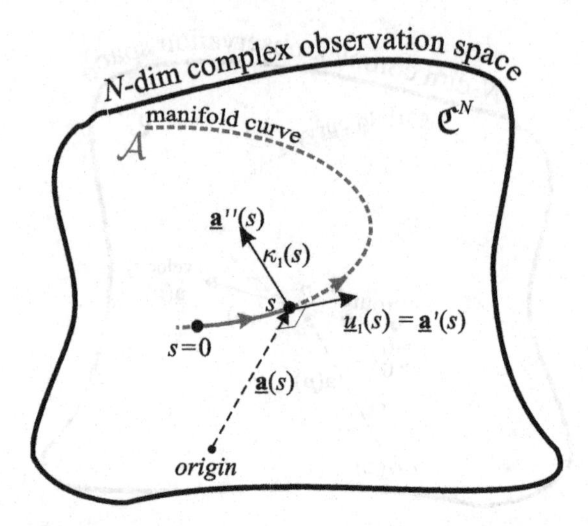

Fig. 2.5 How fast the curve pulls away from the tangent line, at point s, is measured by $\underline{a}''(s)$ with $\|\underline{a}''(s)\| = \kappa_1(s)$.

2.2 Curvatures and Coordinate Vectors in $\mathfrak{C}^N$

We have seen that the arc length s is the invariant parameter of a manifold curve of an array of N sensors. With the array manifold curve embedded in an N-dimensional complex space (or equivalently in a $2N$ real space) and parametrized in terms of the arc length s, it is essential to attach/define to the running point $\underline{a} = \underline{a}(s)$ on the curve a number of *curvatures* as well as a continuous differential and orthonormal system of coordinates $\underline{u}_1(s)$, $\underline{u}_2(s), \ldots, \underline{u}_{2N}(s)$.

2.2.1 *Number of Curvatures and Symmetricity in Linear Arrays*

The curvatures of a space curve are of immense value in differential geometry as, according to the fundamental uniqueness theorem of [2, 17], curvatures uniquely define a space curve expressed in terms of arc length s, except its position in space.

For an array of N sensors (i.e. the manifold vector $\underline{a}(s) \in \mathfrak{C}^N$ may be described by $2N$ real components), at most $2N - 1$ manifold curvatures can be defined. However, if the array has some symmetrical sensors with respect to the array centroid, then the manifold curve is situated wholly in some *subspace* of dimensionality d (within the $2N$ dimensional space). Therefore, in this case up to $d-1$ non-zero curvatures can be defined. In particular, if m denotes the number of sensors in symmetrical pairs then d is given as follows:

$$d = \begin{cases} 2N - m & \text{if } \nexists \text{ sensor at the array centroid} \\ 2N - m - 1 & \text{otherwise} \end{cases} \tag{2.8}$$

This implies that $0 \leq m \leq N$ and according to m, linear sensor arrays are divided into the following broad categories:

(1) Symmetric: $m = N$, i.e. all the sensors occur in symmetric pairs about the origin.
(2) Non-Symmetric: $m < N$
  (a) Partially symmetric: $0 < m < N$, i.e. at least one sensor has a symmetric counterpart about the origin;
  (b) Fully asymmetric: $m = 0$, i.e. no sensor has a symmetrical counterpart about the origin.

Note that a sensor at the origin is taken to be a symmetrical sensor. A representative example from each category is illustrated in Figs. 2.6, 2.7 and 2.8.

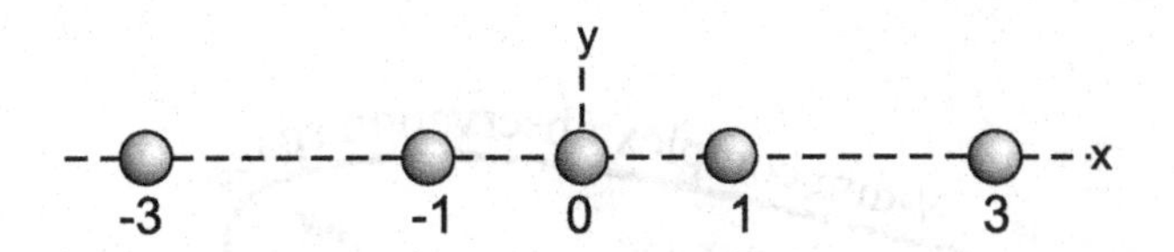

Fig. 2.6 A symmetric linear array of $N = 5$ elements ($m = 5$, $d = 2N - m - 1 = 4$).

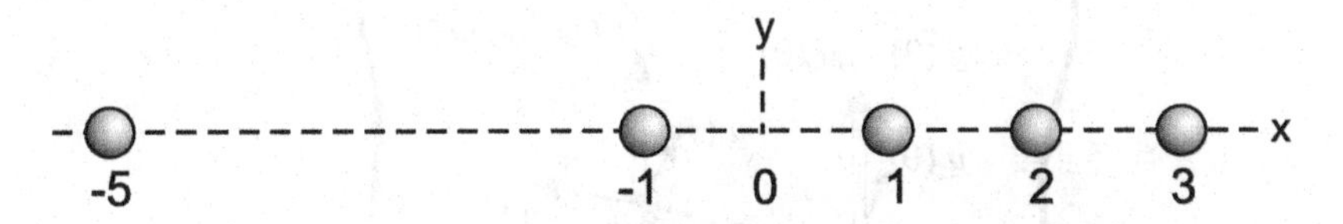

Fig. 2.7 A partially symmetric linear array of $N = 5$ elements ($m = 2$, $d = 2N - m = 8$).

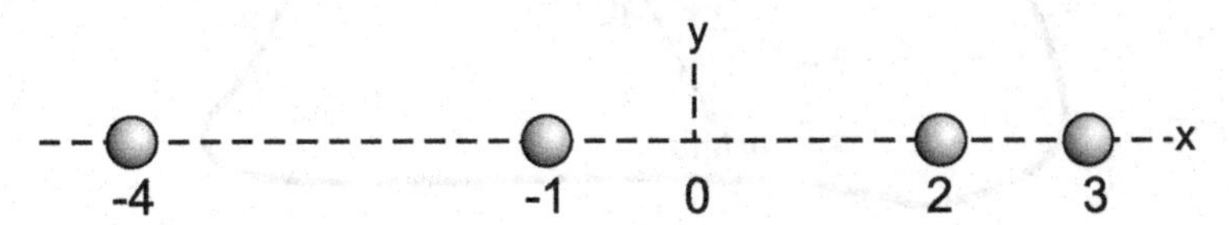

Fig. 2.8 An asymmetric linear array of $N = 4$ elements ($m = 0$, $d = 2N - m = 8$).

2.2.2 *"Moving Frame" and Frame Matrix*

In the previous section we have seen that to the running point $\underline{\mathbf{a}}(s)$ on the curve $\mathcal{A}$ we attach a frame of orthonormal vectors $\underline{u}_1(s)$, $\underline{u}_2(s), \ldots, \underline{u}_{2N}(s)$ together with $d-1$ non-zero curvatures $\kappa_1(s), \ldots, \kappa_{d-1}(s)$ with $N-1 \leq d \leq 2N$.

By ignoring the coordinate vectors with indices greater than d (those corresponding to zero curvatures) a frame of d orthonormal vectors $\underline{u}_1(s)$, $\underline{u}_2(s), \ldots, \underline{u}_d(s)$, can be defined and attached to the running point $\underline{\mathbf{a}}(s)$ on the manifold curve, making $\underline{\mathbf{a}}(s)$ the "origin" of the new coordinate system. This set of coordinate vectors, known as a "moving frame" (see Fig. 2.9), forms the matrix

$$\mathbb{U}(s) = [\underline{u}_1(s), \underline{u}_2(s), \ldots, \underline{u}_d(s)] \in \mathfrak{C}^{N \times d} \tag{2.9}$$

and is derived from a fixed known frame $\mathbb{U}(0)$, (i.e. at $s = 0$ say) by rotation, using the transformation matrix $\mathbb{F}(s) \in \mathfrak{C}^{d \times d}$, i.e.

$$\mathbb{U}(s) = \mathbb{U}(0) \cdot \mathbb{F}(s) \quad \text{where} \quad \mathbb{F}(0) = \mathbb{I}_d \tag{2.10}$$

The matrix $\mathbb{F}(s)$ is a continuous differential real transformation matrix called *frame matrix*. This is a non-singular matrix with its main properties listed in Table 2.1.

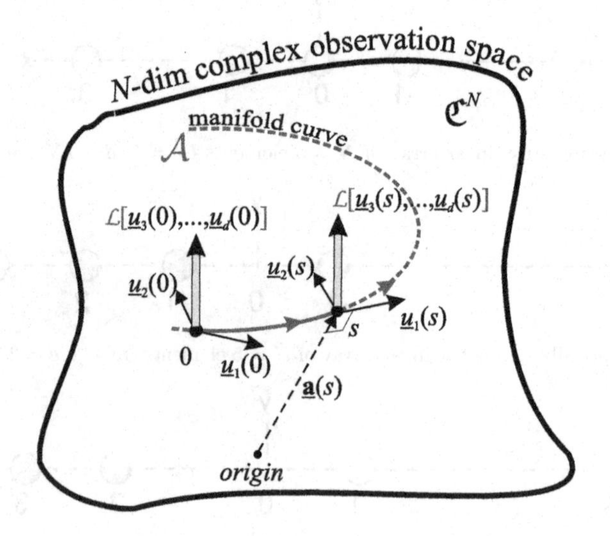

Fig. 2.9 "Moving frame" $\mathbb{U}(0)$ and $\mathbb{U}(s)$.

Table 2.1 Properties of the frame matrix $\mathbb{F}(s)$.

1st property	$\mathbb{F}(s) \in \mathcal{R}^{d\times d}$
2nd property	$\det(\mathbb{F}(s)) = 1$
3rd property	$\mathbb{F}^{-1}(s) = \mathbb{F}^T(s)$
4th property	$\mathbb{F}^T(s) \cdot \mathbb{F}(s) = \mathbb{I}_d$
5th property	$\text{ele}_{ij}(\mathbb{F}) = (-1)^{i+j}\text{ele}_{ji}(\mathbb{F})$
6th property	$\sum_{\substack{i=2 \\ \text{even}}}^{d} \text{ele}_{ii}(\mathbb{F}) = \frac{\text{Tr}(\mathbb{F})-\varepsilon}{2}$ where $\varepsilon = \begin{cases} 0 & \text{if } d \text{ even} \\ 1 & \text{if } d \text{ odd} \end{cases}$

To summarize, the $d \times d$ real matrix $\mathbb{F}(s)$ is known as the "frame matrix" while the set of coordinates represented by the $N \times d$ complex matrix $\mathbb{U}(s)$ is the "moving frame."

2.2.3 *Frame Matrix and Curvatures*

The question is, how the frame matrix $\mathbb{F}(s)$, at the running point s, is related to the curvatures of the manifold attached to this point. The initial part of the mechanism to answer this question is to define the coordinate vectors in $\mathfrak{C}^N$ by exploiting and extending the procedure used in real spaces where a basis of up to N orthonormal vectors is necessary to fully define a curve in $\mathcal{R}^N$ [17]. This extension, which provides the d coordinate vectors in $\mathfrak{C}^N$ blended with the curvatures, is summarized in a structured formation in Table 2.2.

Now let us focus on the $(i+1)$th row of Table 2.2. That is,

$$\underline{u}_{i+1}(s) = \frac{\underline{u}_i'(s) + \kappa_{i-1}.\underline{u}_{i-1}(s)}{\kappa_i} \tag{2.11}$$

By solving Eq. (2.11) with respect to $\underline{u}_i'(s)$, it is obvious that the differentiation of the ith coordinate vector, for $i \geqslant 2$, can be written as

$$\underline{u}_i'(s) = \kappa_i(s)\underline{u}_{i+1}(s) - \kappa_{i-1}(s)\underline{u}_{i-1}(s) \quad \text{for } i \geqslant 2 \tag{2.12}$$

with

$$\underline{u}_1'(s) = \underline{\mathbf{a}}''(s) = \kappa_1(s)\underline{u}_2(s)$$

Note that the coordinate vectors are normalized to unity length.

Equation (2.12) which can be rewritten in a more compact form as

$$\mathbb{U}'(s) = \mathbb{U}(s)\,\mathbb{C}(s) \tag{2.13}$$

Table 2.2 Estimation procedure of coordinate vectors and curvatures of a curve $\mathcal{A}$.

	Coordinate Vector	Curvature
1st	$\underline{u}_1(s) \triangleq \underline{\mathbf{a}}'(s)$	$\kappa_1(s) = \lVert\underline{u}_1'(s)\rVert$
2nd	$\underline{u}_2(s) = \frac{1}{\kappa_1}\underline{u}_1'(s)$	$\kappa_2(s) = \lVert\underline{u}_2'(s) + \kappa_1\underline{u}_1(s)\rVert$
3rd	$\underline{u}_3(s) = \dfrac{\underline{u}_2'(s) + \kappa_1\underline{u}_1(s)}{\kappa_2}$	$\kappa_3(s) = \lVert\underline{u}_3'(s) + \kappa_2\underline{u}_2(s)\rVert$
...	...	...
...	...	...
ith	$\underline{u}_i(s) = \dfrac{\underline{u}_{i-1}'(s) + \kappa_{i-2}\cdot\underline{u}_{i-2}(s)}{\kappa_{i-1}}$	$\kappa_i(s) = \lVert\underline{u}_i'(s) + \kappa_{i-1}\cdot\underline{u}_{i-1}(s)\rVert$
$(i$+1)th	$\underline{u}_{i+1}(s) = \dfrac{\underline{u}_i'(s) + \kappa_{i-1}\cdot\underline{u}_{i-1}(s)}{\kappa_i}$	$\kappa_{i+1}(s) = \lVert\underline{u}_{i+1}'(s) + \kappa_i\cdot\underline{u}_i(s)\rVert$
...	...	...
...	...	...
dth	$\underline{u}_d(s) = \dfrac{\underline{u}_{d-1}'(s) + \kappa_{d-2}\cdot\underline{u}_{d-2}(s)}{\kappa_{d-1}}$	$\kappa_d(s) = 0$

where $\mathbb{U}'(s) = [\underline{u}_1'(s), \underline{u}_2'(s), \ldots, \underline{u}_d'(s)] \in \mathfrak{C}^{N\times d}$, d is the dimensionality of the subspace in which the curve is embedded. The matrix $\mathbb{C}(s) \in \mathfrak{C}^{d\times d}$ is termed the Cartan matrix, which is a real skew-symmetric matrix of the curvatures defined as follows:

$$\mathbb{C}(s) \triangleq \begin{bmatrix} 0, & -\kappa_1(s), & 0, & \ldots, & 0, & 0 \\ \kappa_1(s), & 0, & -\kappa_2(s), & \ldots, & 0, & 0 \\ 0, & \kappa_2(s), & 0, & \ldots, & 0, & 0 \\ \vdots & \vdots & \vdots & \ddots & \vdots & \vdots \\ 0, & 0, & 0, & \ldots, & 0, & -\kappa_{d-1}(s) \\ 0, & 0, & 0, & \ldots, & \kappa_{d-1}(s), & 0 \end{bmatrix} \quad (2.14)$$

Starting with Eq. (2.13) and then using Eq. (2.10), we have

$$\begin{aligned} \mathbb{U}'(s) &= \mathbb{U}(s)\ \mathbb{C}(s) \\ \Rightarrow (\mathbb{U}(0)\ \mathbb{F}(s))' &= \mathbb{U}(0)\ \mathbb{F}(s)\ \mathbb{C}(s) \\ \Rightarrow \mathbb{U}(0)\ \mathbb{F}'(s) &= \mathbb{U}(0)\ \mathbb{F}(s)\ \mathbb{C}(s) \end{aligned}$$

leading to the following expression

$$\mathbb{F}'(s) = \mathbb{F}(s)\ \mathbb{C}(s) \quad (2.15)$$

This is a first order differential equation of the frame matrix $\mathbb{F}(s)$ with initial condition $\mathbb{F}(0) = \mathbb{I}_d$. Therefore, its solution is

$$\mathbb{F}(s) = \text{expm}(s\ \mathbb{C}(s)) \tag{2.16}$$

where expm$(\cdot)$ denotes the matrix exponential. Equation (2.16) provides the relationship between the frame matrix $\mathbb{F}(s)$ at the running point s and the curvatures (Cartan matrix) of the manifold attached to this point, which can also be used to rewrite Eq. (2.10) as follows:

$$\mathbb{U}(s) = \mathbb{U}(0)\ \text{expm}(s\ \mathbb{C}(s)) \tag{2.17}$$

Finally, Eq. (2.15) implies that the Cartan matrix can always be written as a function of the frame matrix $\mathbb{F}(s)$ as follows

$$\mathbb{C}(s) = \mathbb{F}^{-1}(s)\ \mathbb{F}'(s) = \mathbb{F}^T(s)\ \mathbb{F}'(s) \tag{2.18}$$

where the 3rd property of $\mathbb{F}(s)$ given in Table 2.1 has been used. Remember that the Cartan matrix, as a formation of the curvatures, contains all the information about the local behavior and shape of the manifold curve.

2.2.4 *Narrow and Wide Sense Orthogonality*

As stated previously, for a general array manifold curve it is essential to attach, to a running point on the curve, a *continuously differentiable* and *orthonormal* system of coordinates $\mathbb{U}(s)$. The orthonormality of this continuously differentiable "moving" frame relies on the vector continuum $\underline{u}_i(s)$ with constant magnitude and being orthogonal to its tangent. That is,

$$\underline{u}_i(s) \perp \underline{u}_i'(s), \quad \text{i.e.} \quad \underline{u}_i(s)^H \underline{u}_i'(s) = 0$$

This, however, is only true for the broad class of arrays which are symmetric with respect to their centroid, in which case $m = N$. In general, if the array is nonsymmetric ($m < N$), then the orthogonality is not valid although the complex vectors $\underline{u}_i(s)$ have constant (unity) magnitude. In particular, the inner product of a complex vector of constant norm with its derivative is purely imaginary, i.e.

$$\underline{u}_i^H(s)\underline{u}_i'(s) = \text{imaginary} \tag{2.19}$$

This leads us to define orthogonality of two complex vectors of unity magnitude as being "wide-sense" and "narrow-sense" orthogonality, as follows:

Definition 2.1 Wide sense and narrow sense orthogonality of two complex vectors of unity magnitude:

$$\text{“wide” sense orthogonality:} \quad \mathrm{Re}(\underline{u}_i^H(s).\underline{u}_j(s)) = 0, \quad \text{for } i \neq j \tag{2.20}$$

$$\text{“narrow” sense orthogonality:} \quad \underline{u}_i^H(s) \cdot \underline{u}_j(s) = 0, \quad \text{for } i \neq j \tag{2.21}$$

Thus, for a general linear array, that is, for a non-symmetric array, the wide-sense orthogonality should be used, i.e.

$$\mathrm{Re}(\mathbb{U}^H(s) \cdot \mathbb{U}(s)) = \mathbb{I}_d \tag{2.22}$$

This, of course, is simplified to narrow-sense orthogonality $\mathbb{U}^H(s) \cdot \mathbb{U}(s) = \mathbb{I}_d$ for symmetrical arrays.

Using the “wide sense” orthogonality for a general linear array, and starting with Eq. (2.18), we get

$$\mathrm{Re}(\mathbb{U}^H(s) \cdot \mathbb{U}'(s)) = \mathbb{C}(s) \tag{2.23}$$

Indeed

$$\begin{aligned}
\mathbb{C}(s) &= \mathbb{F}^T(s) \cdot \mathbb{F}'(s) = \mathbb{F}^T(s) \cdot \mathbb{I}_d \cdot \mathbb{F}'(s) = \mathbb{F}^T(s) \cdot \underbrace{\mathrm{Re}(\mathbb{U}^H(0) \cdot \mathbb{U}(0))}_{=\mathbb{I}_d} \cdot \mathbb{F}'(s) \\
&= \mathrm{Re}\left(\underbrace{\mathbb{F}^T(s) \cdot \mathbb{U}^H(0)}_{=\mathbb{U}^H(s)} \cdot \underbrace{\mathbb{U}(0)\mathbb{F}'(s)}_{=\mathbb{U}'(s)} \right) = \mathrm{Re}(\mathbb{U}^H(s)\mathbb{U}'(s))
\end{aligned}$$

Furthermore, as it is shown in Appendix 2.7.1, the ith derivative of the manifold vector with respect to the arc length s (denoted by $\underline{\mathbf{a}}'^{(i)}(s)$) is related to the coordinate vector in a simple way as follows:

$$\begin{aligned}
\mathrm{Re}\left(\left(\underline{\mathbf{a}}'^{(i)}(s) \right)^H \cdot \underline{u}_i(s) \right) &= \kappa_1(s) \cdot \kappa_2(s) \cdots \kappa_{i-1}(s) \\
&= \prod_{\ell=1}^{i-1} \kappa_\ell(s)
\end{aligned} \tag{2.24}$$

Based on Eq. (2.24), it is easy to prove that the determinant of the $N \times N$ matrix formed by the first N derivatives of the manifold vector is related to the curvatures by

$$\begin{aligned}
&|\det[\underline{\mathbf{a}}'(s), \underline{\mathbf{a}}''(s), \ldots, \underline{\mathbf{a}}'^{(N)}(s)]| \\
&= \kappa_1^{N-1}(s) \cdot \kappa_2^{N-2}(s) \cdots \kappa_{N-2}^{2}(s) \cdot \kappa_{N-1}(s) \\
&= \prod_{\ell=1}^{N-1} \kappa_\ell^{N-\ell}(s)
\end{aligned} \tag{2.25}$$

2.3 "Hyperhelical" Manifold Curves

In this section, the shape of the array manifold of a linear array of N isotropic sensors is investigated using the theory developed in the previous section. The manifold is parametrized in terms of a directional parameter p with the sensor locations given by the vector $\underline{r}$ in half-wavelengths. The theory is valid for any *regular parametrized differential* manifold curve $\mathcal{A}$ defined as follows:

$$\mathcal{A} \triangleq \{\underline{\mathbf{a}}(p) \in \mathfrak{C}^N, \ \forall p : p \in \Omega_p\} \tag{2.26}$$

where

$$\underline{\mathbf{a}}(p) = \exp(-j(\pi \underline{r} \cos p + \underline{v})) \tag{2.27}$$

where $\underline{r}$ and $\underline{v}$ are two constant vectors and p is a generic parameter (parameter of interest). The main results are presented in the form of two theorems but, firstly, it is easy to show using Eq. (2.3) that the arc length $s(p)$, and rate of change of arc length $\dot{s}(p)$, are given by the following expressions respectively:

$$s(p) = \pi \left\|\underline{r}\right\| (1 - \cos p) \quad \text{i.e.} \quad p(s) = \cos^{-1}\left(1 - \frac{s}{\pi\|\underline{r}\|}\right) \tag{2.28}$$

and

$$\dot{s}(p) = \pi\|\underline{r}\| \sin p \tag{2.29}$$

where the initial condition $s(0) = 0$ has been assumed.

It is worth noting that for a linear array of N sensors with locations $\underline{r}$ (in units of half-wavelengths) the rate of change of the arc length is a non-linear function of the directional parameter p and depends on the norm of the vector of sensor locations. From Eq. (2.29), it can be further deduced that for a small directional increment $\Delta p \triangleq |p_2 - p_1|$, the corresponding change in arc length, to a first order approximation, can be calculated as:

$$\Delta s = |s(p_2) - s(p_1)| \tag{2.30}$$

$$= \pi \|\underline{r}\| |\cos p_2 - \cos p_1| \tag{2.31}$$

$$\approx \pi \|\underline{r}\| \Delta p \sin \breve{p}; \quad \text{for small } \Delta p \tag{2.32}$$

where $\breve{p} \triangleq (p_1 + p_2)/2$. This expression reveals crucial properties of the array manifold of a linear array and provides considerable insight into its resolving power. Consider a linear array and two impinging emitters with a directional separation of Δp. It can easily be seen from Eq. (2.32) that the distance (arclength Δs) between the corresponding manifold (response) vectors in $\mathfrak{C}^N$ is maximum when the emitters are at broadside and minimum when they are at endfire. Furthermore, if a larger array is employed, then the corresponding manifold vectors are further apart in $\mathfrak{C}^N$. These facts are demonstrated by the variations of the rate of change of arc length with azimuth for 5-sensor and 7-sensor standard ULAs of half-wavelength spacings, plotted in Fig. 2.10.

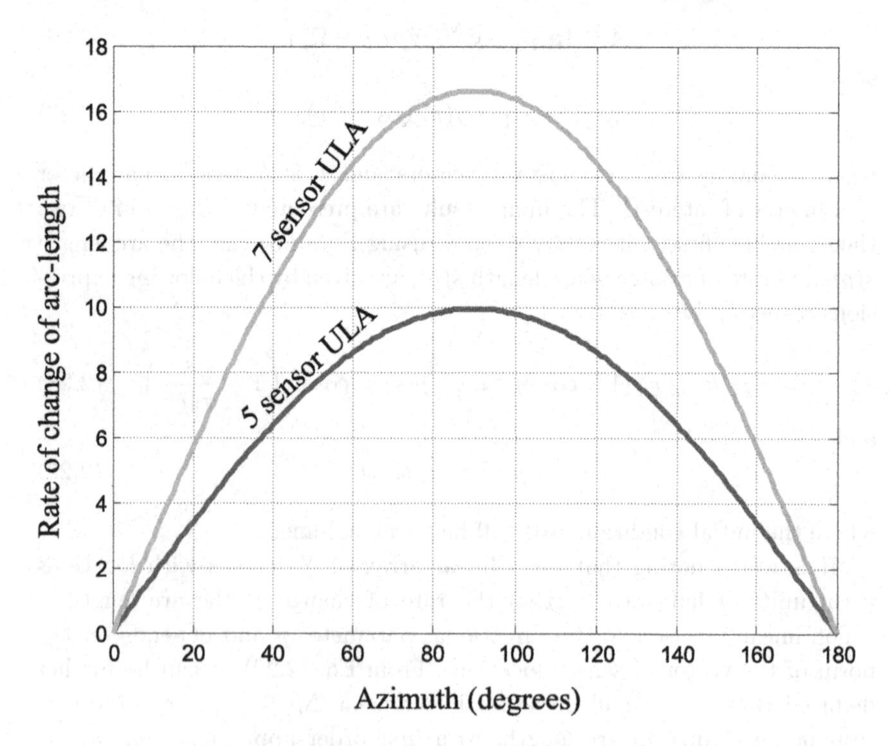

Fig. 2.10 Variations of $\dot{s}(\theta)$ for ULAs of 5 and 7 sensors.

This explains the better resolution of sources at broadside and the better resolving power of the 7-sensor standard ULA, as evidenced by Fig. 2.10. It can therefore be concluded that the rate of change of arc length of the array manifold plays a prominent role in determining the resolving power of the array, as will be further detailed in Chapter 8.

Another important parameter of the array manifold curve is its total length,

$$l_{\mathrm{m}} \triangleq s(\pi) = 2\pi \, \|\underline{r}\| \tag{2.33}$$

which clearly increases in direct proportion to the sensor spacings. As can be seen, the manifold length depends on the number of sensors and their positions. The manifold length is expected to influence the ambiguity properties of a linear array since it is obvious that longer manifolds are more prone to "spurious" parameters/results (manifold vectors that can be expressed as linear combinations of other manifold vectors) unless the arrays are carefully designed.

The array manifold can now be parametrized in terms of its arc length as follows:

$$\mathcal{A} = \{\underline{\mathbf{a}}(s) \in \mathfrak{C}^N, \ \forall s : s \in [0, l_{\mathrm{m}}]\} \tag{2.34}$$

where

$$\underline{\mathbf{a}}(s) \triangleq \exp(j(\underline{\widetilde{r}}s - \pi\underline{r} + \underline{v})) \tag{2.35}$$

and

$$\underline{\widetilde{r}} \triangleq \frac{\underline{r}}{\|\underline{r}\|} \tag{2.36}$$

is the vector of "normalized" sensor positions.

From Eq. (2.35), and bearing in mind that the coordinate vectors are of unit length, the following expressions for the first three manifold curvatures, for instance, may be derived from the first three rows of Table 2.2

$$\begin{aligned}
\underline{u}_1(s) &\triangleq \underline{\mathbf{a}}'(s) = j\underline{\widetilde{r}} \odot \underline{\mathbf{a}}(s) \\
\underline{u}_2(s) &= \frac{1}{\kappa_1}\underline{u}_1'(s) = -\frac{1}{\kappa_1}\underline{\widetilde{r}}^2 \odot \underline{\mathbf{a}}(s) \\
\underline{u}_3(s) &= \frac{1}{\kappa_2}(\underline{u}_2'(s) + \kappa_1\underline{u}_1(s)) = -\frac{j}{\kappa_1\kappa_2}(\underline{\widetilde{r}}^3 - \kappa_1^2\underline{\widetilde{r}}) \odot \underline{\mathbf{a}}(s) \\
\kappa_1(s) &= \|\underline{u}_1'(s)\| = \|\underline{\widetilde{r}}^2\| \\
\kappa_2(s) &= \|\underline{u}_2'(s) + \kappa_1\underline{u}_1(s)\| = \frac{1}{\kappa_1}\|\underline{\widetilde{r}}^3 - \kappa_1^2\underline{\widetilde{r}}\| \\
\kappa_3(s) &= \|\underline{u}_3'(s) + \kappa_2\underline{u}_2(s)\| = \frac{1}{\kappa_1\kappa_2}\|\underline{\widetilde{r}}^4 - (\kappa_1^2 + \kappa_2^2)\underline{\widetilde{r}}^2\|
\end{aligned} \tag{2.37}$$

In general, it can be shown that the ith coordinate vector and curvature of the manifold of a linear array of isotropic sensors can be calculated according to the following theorem.

Theorem 2.1 ***Hyperhelical Manifolds:*** *The manifold of any linear array of N omnidirectional sensors with locations given by $\underline{r}$,*

(a) is a curve of hyperhelical shape lying on a complex N-dimensional sphere with radius $\sqrt{N}$.
(b) The coordinate vectors and curvatures of this hyperhelix depend on the lower order curvatures, the number of sensors and their relative spacing and can be estimated by the following recursive equations:

$$\underline{u}_i(s) = \frac{(j)^i}{\kappa_1 \kappa_2 \cdots \kappa_{i-1}} \sum_{n=1}^{\text{fix}(\frac{i-1}{2})+1} (-1)^{n-1} b_{i-1,n} \, \underline{\widetilde{r}}^{i-2n+2} \odot \underline{\mathbf{a}}(s) \quad (2.38)$$

$$\kappa_i = \frac{1}{\kappa_1 \kappa_2 \cdots \kappa_{i-1}} \left\| \sum_{n=1}^{\text{fix}(\frac{i}{2})+1} (-1)^{n-1} \, b_{i,n} \, \underline{\widetilde{r}}^{i-2n+3} \right\| \quad (2.39)$$

where

$$\begin{cases} \underline{\widetilde{r}} = \frac{\underline{r}}{\|\underline{r}\|} \text{ (normalized sensor positions)} \\ \kappa_1 = \|\underline{\widetilde{r}}^2\| \\ \kappa_{i-1} \neq 0 \\ \text{sum}(\underline{r}) = 0 \text{ (i.e. phase reference = array centroid)} \end{cases}$$

Furthermore, the coefficients $b_{i,n}$ are given by:

$$b_{i,n} = \sum_{m_1=1}^{i-2n+3} \sum_{m_2=2+m_1}^{i-2n+5} \cdots \sum_{m_{n-1}=2+m_{n-2}}^{i-1} \kappa_{m_1}^2 \kappa_{m_2}^2 \cdots \kappa_{m_{n-1}}^2; \quad i, n > 2 \quad (2.40)$$

with

$$\left. \begin{aligned} b_{i,1} &= 1; & i \geqslant 1 \\ b_{i,n} &= \sum_{m=1}^{i-1} \kappa_m^2 & i > 1 \end{aligned} \right\}$$

or recursively by

$$b_{i,n} = b_{i-1,n} + \kappa_{i-1}^2 b_{i-2,n-1}, \quad i > 2, \ n > 1 \quad (2.41)$$

with the initial conditions:

$$\begin{cases} b_{i,1} = 1, & i \geq 1 \\ b_{2,2} = \kappa_1^2 \end{cases} \tag{2.42}$$

Corollary 2.1 *The manifolds of uniform linear arrays are hypercircular arcs (closed hyperhelices).*

The proof of the above theorem can be found in Appendix 2.7.2 (page 54). Note that for notational convenience the second condition (row) of Eq. (2.8) may be ignored by redefining $d \triangleq 2N - m$ with $N \leq d \leq 2N$. This implies that in the special case of an array with a sensor at the array centroid, we attach to the running point $\underline{\mathbf{a}}(s)$ on the curve $\mathcal{A}$ a moving frame of $\underline{u}_1(s)$, $\underline{u}_2(s)$, ..., $\underline{u}_d(s)$ together with $d-2$ non-zero curvatures and one zero curvature ($\kappa_{d-1}(s) = 0$). In this special case Eq. (2.38) cannot be used to calculate the dth coordinate vector $\underline{u}_d(s)$ since the curvature κ_{d-1} vanishes. Using an orthogonalization procedure, the $(N \times 1)$ vector $\underline{u}_d(s)$ can be shown to be given by

$$\underline{u}_d(s) = [0, \ldots, 0, 1, 0, \ldots, 0]^T \tag{2.43}$$

where the non-zero entry is in the same position as the sensor at the centroid in the vector of sensor locations $\underline{\widetilde{r}}$.

2.3.1 *Coordinate Vectors and Array Symmetricity*

A close examination of Eq. (2.39) of Theorem 2.1, reveals some important properties of the manifold of a linear array of isotropic sensors. In particular, unlike the real case, the coordinate vectors of the linear array manifold curve embedded in $\mathfrak{C}^N$ are not in general mutually orthogonal. For example:

$$\left.\begin{aligned} \underline{u}_1^H(s)\underline{u}_2(s) &= \frac{j}{\kappa_1}\underline{1}_N^T\underline{\widetilde{r}}^3 \\ \underline{u}_1^H(s)\underline{u}_3(s) &= 0 \\ \underline{u}_2^H(s)\underline{u}_3(s) &= \frac{j}{\kappa_2}\left(\frac{1}{\kappa_1^2}\underline{1}_N^T\underline{\widetilde{r}}^5 - \underline{1}_N^T\underline{\widetilde{r}}^3\right) \\ \underline{u}_2^H(s)\underline{u}_4(s) &= 0 \end{aligned}\right\} \tag{2.44}$$

It can, however, be seen that some of the coordinate vectors (some columns of matrix $\mathbb{U}(s)$) are mutually orthogonal. Indeed, it can be shown that the odd-indexed coordinate vectors forming the matrix

$$\mathbb{U}_{\text{odd}}(s) = \left[\underline{u}_1(s), \underline{u}_3(s), \underline{u}_5(s), \ldots, \underline{u}_{d_{odd}}(s)\right] \tag{2.45}$$

are mutually orthogonal and that the same holds for the even-indexed coordinate vectors forming the matrix

$$\mathbb{U}_{\text{even}}(s) = \left[\underline{u}_2(s), \underline{u}_4(s), \underline{u}_6(s), \ldots, \underline{u}_{d_{even}}(s)\right] \tag{2.46}$$

where $d_{\text{even}} = \text{fix}(\frac{d}{2})$ and $d_{\text{odd}} = \text{fix}(\frac{d-1}{2}) + 1$. This can be expressed in a compact form as

$$\left.\begin{array}{c} \mathbb{U}_{\text{odd}}^H(s)\mathbb{U}_{\text{odd}}(s) = \mathbb{I}_{d_{\text{odd}}} \\ \mathbb{U}_{\text{even}}^H(s)\mathbb{U}_{\text{even}}(s) = \mathbb{I}_{d_{\text{even}}} \end{array}\right\} \tag{2.47}$$

with

$$\left.\begin{array}{c} \mathbb{U}_{\text{odd}}^H(s)\mathbb{U}_{\text{even}}(s) \neq \mathbb{O}_{d_{\text{odd}} \times d_{\text{even}}} \\ \text{Re}\{\mathbb{U}_{\text{odd}}^H(s)\mathbb{U}_{\text{even}}(s)\} = \mathbb{O}_{d_{\text{odd}} \times d_{\text{even}}} \end{array}\right\} \tag{2.48}$$

Note also that, at broadside (i.e. $p = 90°$), $\mathbb{U}_{\text{odd}}(s)$ is an imaginary matrix while $\mathbb{U}_{\text{even}}(s)$ is a real matrix. In addition, in the special case of "symmetric arrays", the coordinate vectors form a mutually orthogonal frame

$$\text{symmetric arrays:} \quad \mathbb{U}^H(s)\mathbb{U}(s) = \mathbb{I}_N; \tag{2.49}$$

since the sum of odd powers of the sensor locations is equal to zero.

2.3.2 *Evaluating the Curvatures of Uniform Linear Array Manifolds*

The manifold curvatures (and hence the Cartan matrix $\mathbb{C}$) of a linear array of isotropic sensors are constant, or equivalently, independent of parameters s or p. They depend, however, on the relative rather than the absolute sensor spacings. For instance, the manifold of a 3-element linear array with intersensor spacings 1.5 and 1 has the same curvatures with a 3-element linear array with spacings 0.75 and 0.5 (where the spacings are measured in half wavelengths) but the length of its manifold is twice as long.

Using Theorem 2.1, all the curvatures of any linear array can be calculated. For instance, in Table 2.3, the first eight curvatures are tabulated for arrays with 3 to 10 elements chosen from the popular class of uniform linear arrays with half-wavelength spacing. Furthermore, in Fig. 2.11 the first four curvatures are shown versus the number of elements in the array. It is apparent from Fig. 2.11 that the curvatures are monotically decreasing as the number of elements exceeds twice the order of the curvature.

In Fig. 2.12, the curvatures of a 13-element uniform linear array of sensors are seen to follow a fading oscillating pattern.

It is important to point out that, because the manifold has constant principal (first) curvature and lies on a sphere with radius $\sqrt{N}$, a lower limit for κ_1 is

$$\kappa_1 > \frac{1}{\sqrt{N}} \tag{2.50}$$

Table 2.3 The first eight curvatures of the manifold of uniform linear arrays.

	$N=3$	$N=4$	$N=5$	$N=6$	$N=7$	$N=8$	$N=9$	$N=10$
$\|\underline{r}\|$	1.4142	2.2361	3.1623	4.1833	5.2915	6.4807	7.7460	9.0830
κ_1	0.7071	0.6403	0.5801	0.5372	0.5000	0.4693	0.4435	0.4214
κ_2	0	0.1874	0.2058	0.2047	0.1984	0.1908	0.1832	0.1760
κ_3	—	0.2343	0.3430	0.3603	0.3568	0.3469	0.3352	0.3234
κ_4	—	0	0	0.1489	0.1783	0.1880	0.1899	0.1884
κ_5	—	—	—	0.1323	0.2270	0.2554	0.2647	0.2660
κ_6	—	—	—	0	0	0.1201	0.1502	0.1637
κ_7	—	—	—	—	—	0.0895	0.1694	0.1002
κ_8	—	—	—	—	—	0	0	0.0664

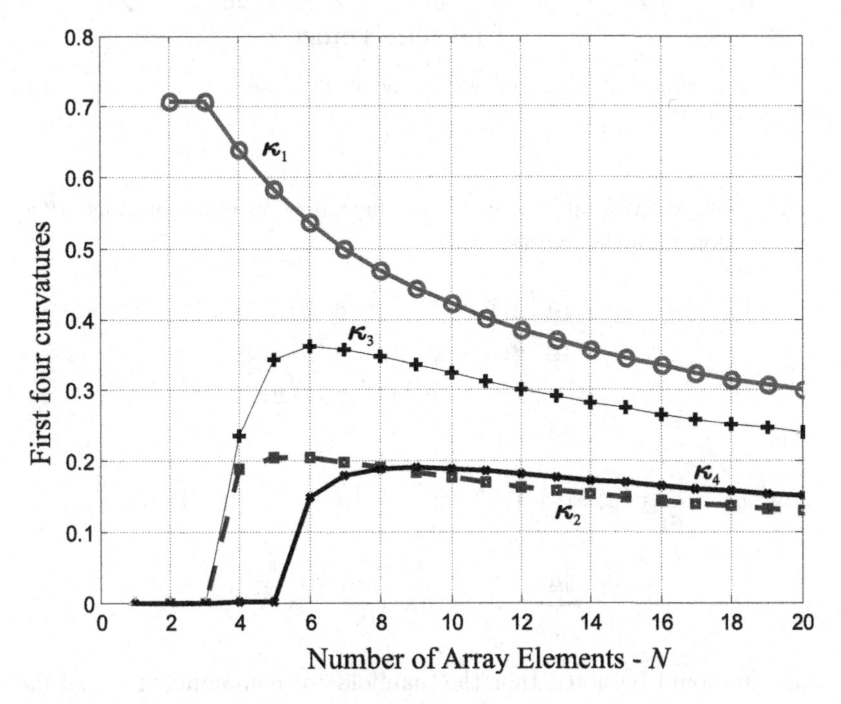

Fig. 2.11 Curvatures of uniform linear arrays as functions of the number of sensors N.

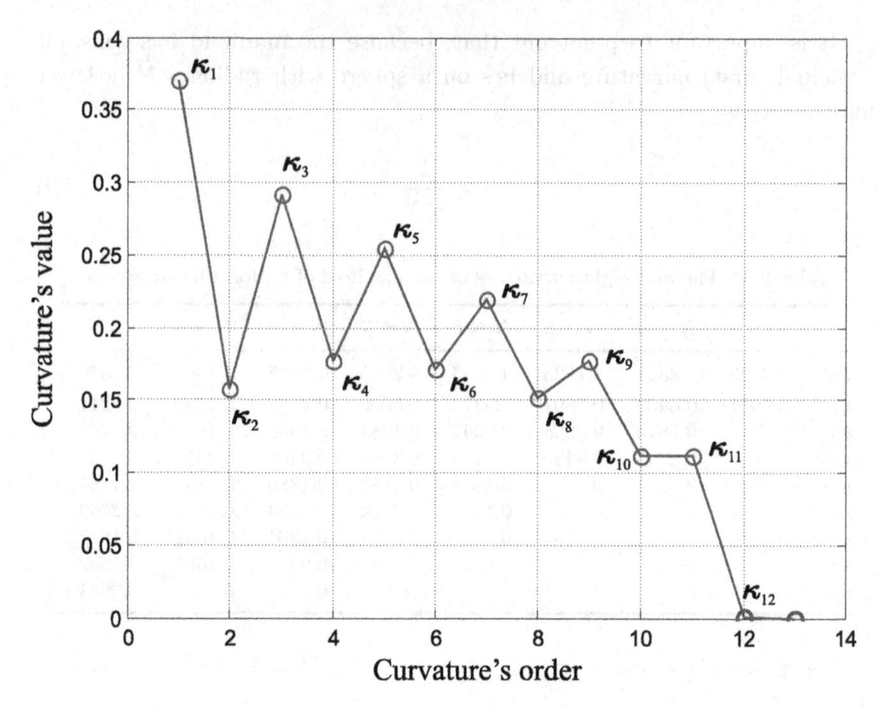

Fig. 2.12 Curvatures of uniform linear array of 13 sensors with half-wavelength intersensor spacing.

The above relation also stems from the following inner product $\underline{\mathbf{a}}^H \underline{u}_2$ in conjunction with the expression

$$\begin{aligned}(\underline{\mathbf{a}}^H \underline{u}_1)' &= \underline{\mathbf{a}}'^H \underline{u}_1 + \underline{\mathbf{a}}^H \underline{u}_1' \\ \Rightarrow (\underline{\mathbf{a}}^H \underline{u}_1)' &= \underline{u}_1^H \underline{u}_1 + \underline{\mathbf{a}}^H \underline{u}_1' \\ \Rightarrow \underline{\mathbf{a}}^H \underline{u}_1' &= (\underline{\mathbf{a}}^H \underline{u}_1)' - \underline{u}_1^H \underline{u}_1\end{aligned}$$

That is,

$$\underline{\mathbf{a}}^H \underline{u}_2 = \frac{1}{\kappa_1} \underline{\mathbf{a}}^H \underline{u}_1' = \frac{1}{\kappa_1} \left((\underline{\mathbf{a}}^H \underline{u}_1)' - \underline{u}_1^H \underline{u}_1 \right) = \frac{1}{\kappa_1}(0 - 1) = -\frac{1}{\kappa_1}$$

$$\Rightarrow \underbrace{\|\underline{\mathbf{a}}\|}_{=\sqrt{N}} . \underbrace{\|\underline{u}_2\|}_{=1} > \frac{1}{\kappa_1} \Rightarrow \kappa_1 > \frac{1}{\sqrt{N}}$$

Finally, it should be noted that the manifolds of non-omnidirectional linear arrays are not hyperhelices. The first curvature of the manifold of such arrays depends both on the relative spacing and the first two derivatives of each elemental pattern $\gamma_k(p)$. In the case of the manifold of a 4-element uniform linear array with non-isotropic, but identical, sensors with directional pattern $\gamma_k(p) = \cos p$, the variation of the first curvature with the azimuth direction is shown in Fig. 2.13. The constant first curvature of the corresponding isotropic array is also shown in the same figure. It is seen that there exist two bearings where the first curvature becomes equal to zero. From the variation of the first curvature, it is possible to deduce the geometrical object to which the manifold best fits. In the case considered, it is apparent that the effect of the sinusoidal elemental pattern is to deform the hyperhelix to a geometrical figure resembling an "eight" with a double point at the origin of the coordinates of $\mathfrak{C}^N$.

2.4 The Manifold Length and Number of Windings (or Half Windings)

It is noted that the manifolds of uniform linear arrays with an odd number of sensors and half-wavelength spacing consist of one winding (round) since the manifold vectors corresponding to the manifold boundaries coincide (i.e. $\underline{a}(0°) = \underline{a}(180°)$). In addition, the manifolds of uniform linear arrays with an even number of sensors and half wavelength spacing, consist of one half winding (half round) since the manifold vectors corresponding to the manifold boundaries are opposite (i.e. $\underline{a}(0°) = -\underline{a}(180°)$). This situation has forced the separation of a hyperhelix corresponding to a general array with an odd number of sensors into consecutive equal arcs of one winding and a hyperhelix corresponding to an array with an even number of sensors into consecutive equal arcs of half-winding.

The following theorem is concerned with the estimation of the number of windings (N odd) and the number of half windings (N even) existing in the manifold of an arbitrary linear array.

Theorem 2.2 ***Windings of a Hyperhelix:*** *The number of windings (N odd), or the number of half windings (N even), of the hyperhelix of Eq. (2.26) is*

$$n_w = \frac{\overbrace{2\pi\,\|\underline{r}\|}^{\text{manifold length}=l_{\mathrm{m}}}}{l_w} \tag{2.51}$$

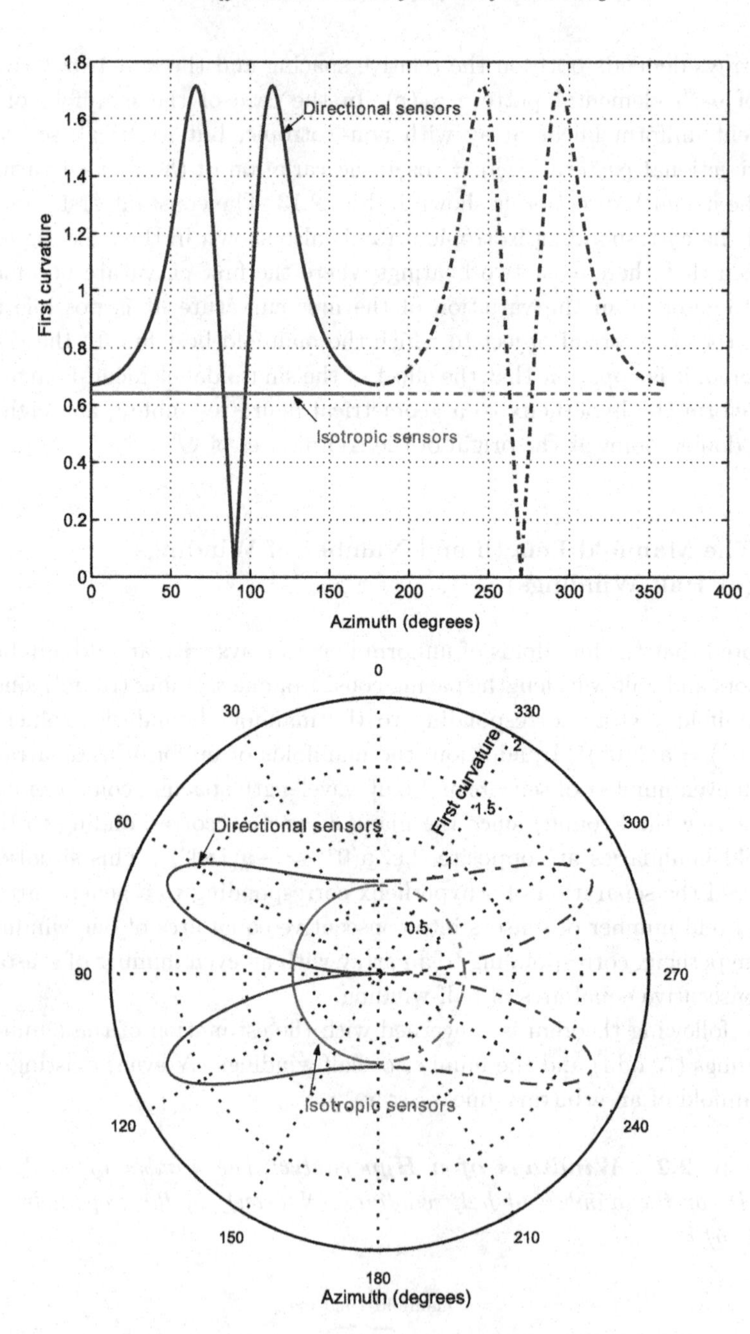

Fig. 2.13 The principal curvature as a function of azimuth (top) and its polar representation (κ, p) (bottom) for a 4-element ULA of directional sensors — $\gamma_k(p) = \cos p$.

where l_w is the positive number that corresponds to the $(N-1)$th root of the function

$$\xi_w(s) = \mathrm{Tr}\left\{\mathbb{C}\,\mathrm{expm}(s\,\mathbb{C})\right\} \tag{2.52}$$

with $\mathbb{C}$ denoting the Cartan matrix.

The independence of the curvatures from the absolute spacings (and hence the independence of the winding length from the absolute spacings) in conjunction with the dependence of the manifold length on the absolute spacings, reveals that two linear arrays with sensors at positions given by the vectors $\underline{r}^{(1)}$ and $\underline{r}^{(2)}$ with $\underline{r}^{(1)} = q\underline{r}^{(2)}$, where q is a scalar constant,

- have manifolds that fit upon each other, and
- the associated number of windings (or half windings) satisfy the following relationship

$$\frac{n_w^{(1)}}{n_w^{(2)}} = q \tag{2.53}$$

It is worth noting that the definition of half winding is restricted to linear arrays with an even number of sensors and

$$\begin{cases} \underline{\mathbf{a}}(0) = \underline{\mathbf{a}}(l_w) \\ \mathbb{U}(0) = \mathbb{U}(l_w) \\ \mathbb{F}(0) = \mathbb{I}_d = \mathbb{F}(l_w) \\ \qquad\quad \Rightarrow \mathrm{Tr}(\mathbb{F}(0)) = \mathrm{Tr}(\mathbb{F}(l_w)) = d \end{cases}$$

where l_w represents the arc length of half winding. For an array with an odd number of sensors, the function ξ_w (Eq. (2.52)) does not generally have a root at the point on the manifold that corresponds to half the length of one manifold winding.

Similarly, the definition of one winding is restricted to linear arrays with an odd number of sensors and

$$\begin{cases} \underline{\mathbf{a}}(0) = -\underline{\mathbf{a}}(l_w) \\ \mathbb{U}(0) = -\mathbb{U}(l_w) \\ \mathbb{F}(0) = \mathbb{I}_d = -\mathbb{F}(l_w) \\ \qquad\quad \Rightarrow \mathrm{Tr}(\mathbb{F}(0)) = -\mathrm{Tr}(\mathbb{F}(l_w)) = d \end{cases}$$

where l_w represents the arc length of one winding. For an array with an even number of sensors, the function ξ_w does not generally have a root at the point on the manifold that corresponds to twice the length of a

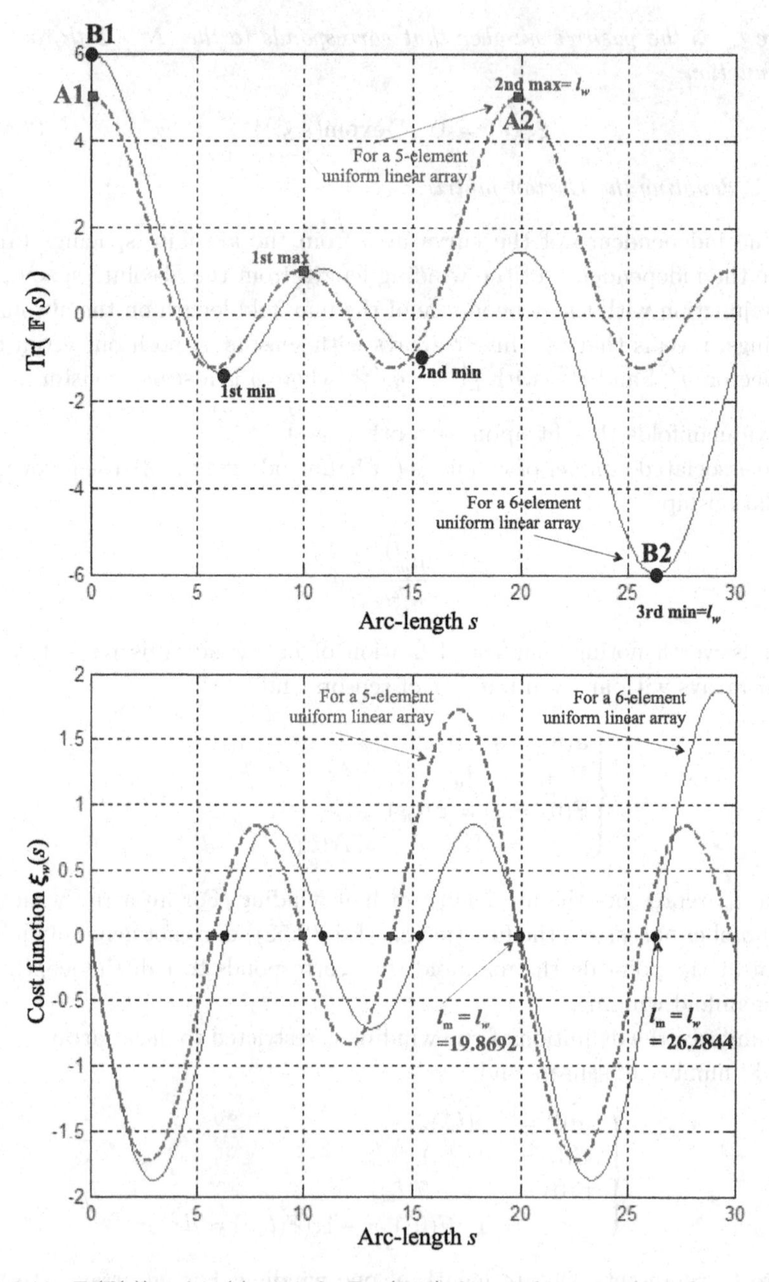

Fig. 2.14 Tr($\mathbb{F}(s)$) and $\xi_w(s)$, given by Eq. (2.52), for a 5-element and a 6-element ULA of half wavelength spacing. Note that $A1 = 5 = A2$ and $B1 = 6 = -B2$.

half-manifold winding. Figure 2.14 provides a representative example for two arrays – with even and odd number of sensors.

2.5 The Concept of "Inclination" of the Manifold

In the previous section it has been shown that the shape of the manifold of a general linear array of omnidirectional sensors is a complex circular hyperhelix lying on the complex N-dimensional sphere with radius $\sqrt{N}$. In this section, an attempt is made to identify the orientation of the array manifold (hyperhelix) on the complex N-dimensional space. Therefore, it is necessary to establish a measure of the orientation of the manifold curve in $\mathfrak{C}^N$ using the concept of *array inclination angle*. This is intuitively defined as follows:

Definition 2.2 Inclination Angle: The inclination angle ζ_{inc} of the manifold of a linear array is the angle formed between any manifold vector $\underline{\mathbf{a}}(s)$ and the subspace $\mathcal{L}\left[\mathbb{U}_{\text{sub}}(s)\right]$, i.e. the subspace spanned by the columns of the matrix

$$\mathbb{U}_{\text{sub}}(s) \triangleq \left[\underline{u}_2(s), \underline{u}_4(s), \underline{u}_6(s), \ldots, \underline{u}_{upto\to N}(s)\right],$$

that is,

$$\zeta_{inc} = \measuredangle\left(\underline{\mathbf{a}}(s), \mathcal{L}\left[\mathbb{U}_{\text{sub}}(s)\right]\right)$$

or equivalently

$$\cos\zeta_{inc} = \sqrt{\frac{1}{N}\ \underline{\mathbf{a}}^H(s)\ \mathbb{P}_{\mathbb{U}_{\text{sub}}}\underline{\mathbf{a}}(s)} \tag{2.54}$$

where $\mathbb{P}_{\mathbb{U}_{\text{sub}}}$ is the projection operator onto $\mathcal{L}\left[\mathbb{U}_{\text{sub}}(s)\right]$.

Having defined the matrix $\mathbb{U}_{\text{sub}}(s)$ as a submatrix of $\mathbb{U}_{\text{even}}(s)$ and because of the orthonormality of columns of the matrix $\mathbb{U}_{\text{even}}(s)$ (see Eq. (2.47)) it is apparent that Eq. (2.54) can be rewritten as follows:

$$\cos\zeta_{inc} = \sqrt{\frac{1}{N}\sum_{i=even}^{N}\left|\underline{\mathbf{a}}^H(s)\ \underline{u}_i(s)\right|^2} \tag{2.55}$$

It is intuitively expected that the inclination angle should increase with the degree of non-symmetry of the array, at least in the neighborhood in which the array is nearly symmetrical. This can, in fact, be verified from Fig. 2.15 in which the inclination of the manifolds for a number of

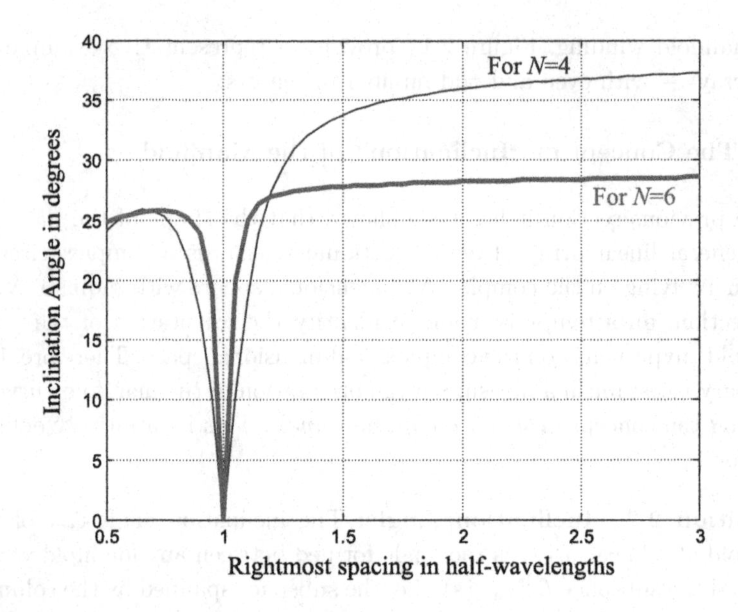

Fig. 2.15 Inclination angle for a number of linear arrays with 4 and 6 sensors.

different arrays resulting from uniform linear arrays with 4 and 6 elements, of half-wavelength spacing, by varying the rightmost (leftmost) spacing from 0.5 half-wavelengths to 3 half-wavelengths is shown.

2.6 The Manifold-Radii Vector

Equation (2.55) shows that the inclination angle is directly related to the inner products of a manifold vector $\underline{\mathbf{a}}^H(s)$ with the coordinate vectors $\underline{u}_2$, $\underline{u}_4, \underline{u}_6, \ldots$, etc. This demonstrates that the relationship between $\underline{\mathbf{a}}^H(s)$ and its associated coordinate vectors $\underline{u}_i(s), \forall i$ is very important.

For example, the relationships between $\underline{\mathbf{a}}^H(s)$ and the first four coordinate vectors, at a point, are:

$$\left.\begin{aligned}
\underline{\mathbf{a}}^H(s)\underline{u}_1(s) &= j\underline{1}_N^T\underline{\widetilde{r}} \\
\underline{\mathbf{a}}^H(s)\underline{u}_2(s) &= -\frac{1}{\kappa_1} \\
\underline{\mathbf{a}}^H(s)\underline{u}_3(s) &= \frac{j}{\kappa_1\kappa_2}\left(\underline{1}_N^T\underline{\widetilde{r}}^3 - \kappa_1^2\underline{1}_N^T\underline{\widetilde{r}}\right) \\
\underline{\mathbf{a}}^H(s)\underline{u}_4(s) &= -\frac{\kappa_2}{\kappa_1\kappa_3}
\end{aligned}\right\} \qquad (2.56)$$

It is easy to see that the orthogonality between a manifold vector and the corresponding tangent vector $\underline{u}_1(s)$ at every point is guaranteed if the phase reference is taken at the array centroid so that

$$\text{sum}\,(\underline{\widetilde{r}}) = 0 \text{ or, equivalent, } \underline{1}_N^T \underline{\widetilde{r}} = 0 \tag{2.57}$$

If the array is also symmetric, then at every point along the manifold, the corresponding response vector is orthogonal to the odd-indexed coordinate vectors.

To provide a more general expression of the inner product

$$\underline{\mathbf{a}}^H(s)\underline{u}_i(s); \quad \forall i$$

the following definition is essential.

Definition 2.3 **Manifold-Radii Vector:** For an N sensor linear array with m sensors in symmetrical pairs the vector

$$\underline{R} = \begin{cases} [0, -R_2, 0, -R_4, 0, \ldots, 0, -R_d]^T \\ \quad \text{if } \nexists \text{ sensor at the array centroid} \\ [0, -R_2, 0, -R_4, 0, \ldots, 0, -R_{d-1}, 1]^T \\ \quad \text{otherwise} \end{cases} \tag{2.58}$$

is defined as the array manifold-radii vector having elements the manifold radii R_i (for $i =$ even) with

$$R_2 = \frac{1}{\kappa_1} \quad \text{and} \quad R_i = \frac{\prod_{n=\text{even}}^{i-2} \kappa_n}{\prod_{n=\text{odd}}^{i-1} \kappa_n} \quad \text{for } i > 2 \tag{2.59}$$

Lemma 2.1 *The inner product $\underline{\mathbf{a}}^H(s)\underline{u}_i(s), \forall i$: For a linear array of N sensors the inner products of a manifold vector $\underline{\mathbf{a}}(s)$ with its d coordinate vectors $\underline{u}_i(s)$ are constant and given by the following expressions:*

$$\underline{\mathbf{a}}^H(s)\underline{u}_i(s) = -R_i \quad \textit{for } i \textit{ even} \tag{2.60}$$

$$\text{Re}(\underline{\mathbf{a}}^H(s)\underline{u}_i(s)) = \begin{cases} 0 & \textit{for } i \neq d \textit{ and odd} \\ 1 & \textit{for } i = d \textit{ and odd} \end{cases}$$

or, in a more compact form, as

$$\underline{\mathbf{a}}^H(s)\mathbb{U}_{\text{even}}(s) = [-R_2, -R_4, \ldots, -R_{d_{\text{even}}}] \tag{2.61}$$

$$\text{Re}\left(\underline{\mathbf{a}}^H(s)\mathbb{U}(s)\right) = \underline{R}^T$$

Note that the inner product of a manifold vector and the corresponding normal vector (or second coordinate vector) is the negative of the radius of curvature (i.e. $-1/\kappa_1$) and hence the other interior products can be considered as higher order "manifold radii." Using the above lemma the inclination angle of Eq. (2.55) can be rewritten as

$$\zeta_{inc} = \arccos\left(\frac{\|\underline{R}_{\text{sub}}\|}{\sqrt{N}}\right) = \text{constant} \tag{2.62}$$

revealing that the inclination is constant, depending only on the norm of a sub-vector of the manifold radii vector and the number of sensors, where $\underline{R}_{\text{sub}}$ is the vector formed by the first N elements of vector $\underline{R}$.

Definition 2.4 Dual arrays: The dual array of a linear array, non-symmetrical with respect to the array centroid, is the array with sensors having positions given by $\underline{r}_i^{\text{dual}} = -\underline{r}_{N\text{-}i+1}$ for $i = 1, \ldots, N$.

Using the above definition it can be seen that, although the manifolds of two dual arrays are not identical, they have the same curvatures and length. Thus, apart from the fact that they have opposite inclinations, they have the same differential geometry properties.

Based on the above concepts the array manifold vector of Eq. (2.27) or (2.35) can be written as

$$\underline{\mathbf{a}}(s) = \mathbb{U}(s)\,\underline{R} = \mathbb{U}(0)\,\mathbb{F}(s)\underline{R} \quad \text{where} \quad \mathbb{F}(s) = \text{expm}(s\mathbb{C}) \tag{2.63}$$

or, by choosing the N coordinate vectors (i.e. N columns of $\mathbb{U}(0)$) at $s = 0$ to be the standard orthonormal basis in $\mathcal{R}^N$, i.e. setting $\mathbb{U}(0) = \mathbb{I}_N$, and expressing s as a function of p using Eq. (2.28),

$$\underline{\mathbf{a}}(p) = \text{expm}(\pi\,\|\underline{r}\|\,(1 - \cos p)\,\mathbb{C})\,\underline{R} \tag{2.64}$$

where $p = 0$ ($s = 0$) is taken along the array axis. Equation (2.64) is associated with a real N-dimensional hyperhelix having the same length and identical curvatures with those of the complex N-dimensional manifold and this is formally described below.

Theorem 2.3 ***Real N-dimensional Hyperhelix of Symmetric Linear Arrays:*** *For a symmetric linear array of N sensors with locations $\underline{r}$, manifold radii vector $\underline{R}$ and Cartan matrix $\mathbb{C}$, there is a* real *N-dimensional hyperhelix described as the locus of the following* *real vector*

$$\underline{\mathbf{a}}_{\text{real}}(p) = \text{expm}(\pi \|\underline{r}\|(1-\cos p)\ \mathbb{C})\underline{R} \in \mathcal{R}^N \tag{2.65}$$

having differential geometry properties equivalent to those of the complex *N-dimensional manifold of the array.*

The vector $\underline{\mathbf{a}}_{\text{real}}(p)$ *can be regarded as the real representation of the manifold of a symmetric array in* $\mathcal{R}^N$.

Thus, the manifolds of symmetrical arrays are shown to admit real representation. This can be seen intuitively from the fact that the components of a manifold vector of symmetrical arrays exist in conjugate pairs. In addition, in the absence of weighting, such arrays have purely real array patterns.

Example 2.1 3-element linear array: The real N-dimensional curve having the same differential geometry properties as the manifold of a 3-element uniform linear array, is a circle. In fact, for $N = 3$, Eq. (2.65) of Theorem 2.3 takes a closed analytical form. The second curvature κ_2 vanishes and the Cartan and Frame matrices are expressed as

$$\mathbb{C} = \begin{bmatrix} 0, & -\kappa_1, & 0 \\ \kappa_1, & 0, & 0 \\ 0, & 0, & 0 \end{bmatrix} \tag{2.66}$$

$$\mathbb{F}(s) = \text{expm}(s\,\mathbb{C}) = \begin{bmatrix} \cos(\kappa_1 s), & \sin(\kappa_1 s), & 0 \\ -\sin(\kappa_1 s), & \cos(\kappa_1 s), & 0 \\ 0, & 0, & 1 \end{bmatrix} \tag{2.67}$$

In addition, the vector $\underline{R}$ is

$$\underline{R} = \left[0, -\frac{1}{\kappa_1}, 1\right]^T \tag{2.68}$$

By applying Eq. (2.65) of Theorem 2.3, the real curve with the same differential geometry properties with the manifold of a 3-element uniform linear array, is given by the real vector continuum

$$\underline{\mathbf{a}}_{\text{real}}(s) = \left[-\frac{1}{\kappa_1}\sin(\kappa_1 s), \quad \frac{1}{\kappa_1}\cos(\kappa_1 s), \quad 1\right]^T \tag{2.69}$$

or, as a function of p,

$$\underline{\mathbf{a}}_{\text{real}}(p) = \left[-\frac{1}{\kappa_1}\sin(\kappa_1 \pi \|\underline{r}\|(1-\cos p)), \quad \frac{1}{\kappa_1}\cos(\kappa_1 \pi \|\underline{r}\|(1-\cos p)), \quad 1\right]^T$$

while the conventional complex array manifold vector of this array is

$$\underline{\mathbf{a}}(p) = \left[\exp\left(j\pi \left\|\underline{r}\right\| \cos p\right), \quad 1, \quad \exp\left(-j\pi \left\|\underline{r}\right\| \cos p\right)\right]^T \tag{2.70}$$

It is clear that Eq. (2.69) is the equation of a closed circle (assuming half-wavelength spacing) with radius $1/\kappa_1$ for $-\|\underline{r}\|\pi \leq s \leq \|\underline{r}\|\pi$. If the two spacings of the above array become unequal, κ_2 does not vanish and the manifold becomes a helix.

Example 2.2 MUSIC Algorithm: To test the proposed results and expressions as well as to demonstrate their use in a DF problem, the MUSIC algorithm is employed, expressed in terms of the curvatures and coordinate vectors. Indeed, if $\mathbb{P}_n$ is the projection operator on to the noise subspace spanned by the noise-level eigenvectors of the array covariance matrix, the cost function of MUSIC can be expressed as a function of the differential geometry parameters of the array manifold as follows:

$$\begin{aligned}
\xi_{\text{MUSIC}}(s) &= \underline{\mathbf{a}}^H(s)\mathbb{P}_n\underline{\mathbf{a}}(s) \\
&= \underbrace{\underline{R}^T\mathbb{F}^T(s)\mathbb{U}^H(0)}_{=\underline{\mathbf{a}}^H(s)}\mathbb{P}_n\underbrace{\mathbb{U}(0)\mathbb{F}(s)\underline{R}}_{\triangleq\underline{\mathbf{a}}(s)} \\
&= \operatorname{Tr}\left(\underbrace{\mathbb{U}^H(0)\mathbb{P}_n\mathbb{U}(0)}_{=\overline{\mathbb{P}}_n}\;\underbrace{\overbrace{\mathbb{F}(s)\underline{R}}^{\triangleq\underline{v}(s)}\overbrace{\underline{R}^T\mathbb{F}(s)}^{=\underline{v}^T(s)}}_{=\mathbb{V}(s)}\right) \\
&= \operatorname{Tr}\left(\overline{\mathbb{P}}_n\mathbb{V}(s)\right)
\end{aligned} \tag{2.71}$$

The matrix $\overline{\mathbb{P}}_n$ is a constant $d \times d$ Hermitian matrix corresponding to a transformation of the second-order statistics of the data collected and is termed the *orientation of the noise subspace.* The matrix $\mathbb{V}(s)$ is a real $d \times d$ symmetric matrix depending exclusively on the curvatures and acts as a differential geometry operator on the suitably transformed data. If the array is symmetric it is apparent that the same expression holds but the matrices $\mathbb{V}(s)$ and $\overline{\mathbb{P}}_n$ are $N \times N$ dimensional real matrices and $\underline{v}(s) = \underline{\mathbf{a}}_{\text{real}}(s)$. Hence, there is a significant reduction in the dimensionality.

In Fig. 2.16, the differential geometry version of MUSIC is simulated for a seven element uniform linear array which operates in the presence of signals with directions of arrival 35°, 120°, and 125°. This figure illustrates that the proper estimation of the three sources is possible by expressing the

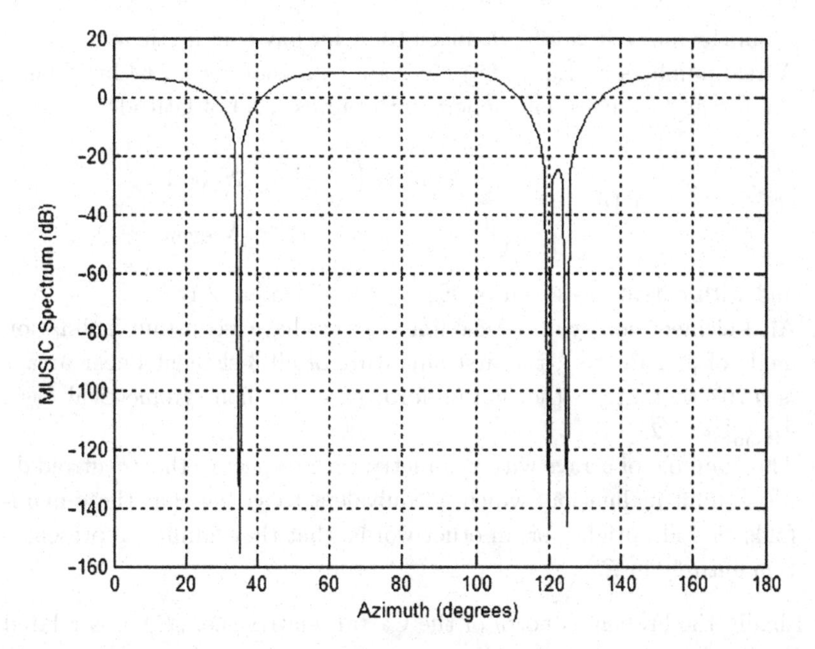

Fig. 2.16 MUSIC spectrum based on Eq. (2.71) for a ULA of 7 sensors operating in the presence of 3 emitting sources.

MUSIC cost function in terms of the differential geometry properties of the manifold.

Deductions

(1) Setting $\|\underline{\mathbf{a}}(s)\| = \sqrt{N}$ and using the wide sense orthonormality of the coordinate vectors, the following relation is derived

$$\|\text{expm}(s\,\mathbb{C})\underline{R}\| = \sqrt{N} \tag{2.72}$$

By setting $s = 0$, it is obvious that

$$\|\underline{R}\| = \sqrt{N} \tag{2.73}$$

(2) If $\underline{R}_{\text{sub}}$ is the vector formed by the first N components of the manifold radii vector, then

$$\|\underline{R}_{\text{sub}}\| \leq \|\underline{R}\| = \sqrt{N} \tag{2.74}$$

However, for symmetrical arrays, $\underline{R}_{\text{sub}} = \underline{R}$, therefore,

$$\text{symmetric arrays: } \|\underline{R}_{\text{sub}}\| = \sqrt{N} \tag{2.75}$$

From the above it can be deduced that, for any symmetric sequence of N real numbers $\underline{r}_i$, Eq. (2.65) gives an expression for a real hyperhelix (on a real N-dimensional sphere with radius $\sqrt{N}$) of latitude

$$\text{latitude} = \begin{cases} \frac{\pi}{2} - \arccos\left(\frac{1}{\sqrt{N}}\right) & N \text{ odd} \\ 0 & N \text{ even} \end{cases}$$

and with curvatures given by Eq. (2.39), Theorem 2.1.

(3) All 3-element non-symmetrical linear arrays have a constant inclination angle of 35.26°. Also, the first curvature of all 3-element linear arrays is 0.707 so that, in the non-uniform case, i.e. non-symmetrical case, $\|\underline{R}_{\text{sub}}\| = \sqrt{2}$.

(4) The manifold of arrays with symmetry with respect to their centroid do not exhibit inclination, which is equivalent to saying that these manifolds "stand upright" or, in other words, that they admit a representation entirely in $\mathcal{R}^N$.

Finally the Frobenius norm of the Cartan matrix (i.e. $\|\mathbb{C}\|_F$) is related to the array symmetricity as follows:

$$\begin{array}{ll} \text{(i)}\ \ \|\mathbb{C}\|_F = 1 & \text{if array} = \text{symmetric} \\ \text{(ii)}\ \|\mathbb{C}\|_F = \sqrt{2} & \text{if array} = \text{fully asymmetric} \\ \text{(iii)}\ 1 < \|\mathbb{C}\|_F < \sqrt{2} & \text{if array} = \text{partially symmetric} \end{array} \tag{2.76}$$

and, at this point, it is convenient to summarize the special properties of the manifolds of symmetric linear arrays:

Symmetric Linear Arrays of N Sensors

- $\text{sum}(\underline{r}^n) = 0$, for n odd
- $m = N$ (with $d \triangleq 2N - m = N$)
- $(\text{No. of } \kappa_i \neq 0) = \begin{cases} N-1 & \text{if} \nexists \text{ sensor at the array centroid} \\ N-2 & \text{otherwise} \end{cases}$
- inclination angle: $\zeta_{inc} = 0^\circ$
- $\mathbb{U}^H(s)\mathbb{U}(s) = \mathbb{I}_N$ (narrow sense orthogonality)
- $\underline{\mathbf{a}}^H(s)\mathbb{U}(s) = \underline{R}^T$
- $\underline{R}_{\text{sub}} = \underline{R}$
- $\underline{\mathbf{a}}_{\text{real}}(p) = \text{expm}(\pi\,\|\underline{r}\|\,(1 - \cos p)\ \mathbb{C})\,\underline{R} \in \mathcal{R}^N$

(2.77)

2.7 Appendices

2.7.1 *Proof of Eq. (2.24)*

The proof will be done by induction. For $i = 2$, we have

$$\underline{\mathbf{a}}''(s)^H \underline{u}_2(s) = \underbrace{\kappa_1 \underline{u}_2(s)^H}_{\underline{\mathbf{a}}''(s)^H} \underline{u}_2(s) = \kappa_1(s) \tag{2.78}$$

Assuming that

$$\text{Re}\left\{\underline{\mathbf{a}}'^{(i-1)}(s)^H \underline{u}_{i-1}(s)\right\} = \kappa_1(s)\kappa_2(s)\cdots\kappa_{i-2}(s) \tag{2.79}$$

we have to prove that

$$\text{Re}\left\{\underline{\mathbf{a}}'^{(i)}(s)^H \underline{u}_i(s)\right\} = \kappa_1(s)\kappa_2(s)\cdots\kappa_{i-1}(s) \tag{2.80}$$

For convenience let us drop the parameter s. The argument of the first part of Eq. (2.80) can be written as follows:

$$\begin{aligned}
&(\underline{\mathbf{a}}'^{(i)})^H \underline{u}_i \\
&= (\underline{\mathbf{a}}'^{(i-1)})'^H \underline{u}_i \\
&= \left((\underline{\mathbf{a}}'^{(i-1)})^H \underline{u}_i\right)' - (\underline{\mathbf{a}}'^{(i-1)})^H \underline{u}_i' \\
&= \left((\underline{\mathbf{a}}'^{(i-1)})^H \underline{u}_i\right)' - (\underline{\mathbf{a}}'^{(i-1)})^H \underbrace{(\kappa_i \underline{u}_{i+1} - \kappa_{i-1}\underline{u}_{i-1})}_{=\underline{u}_i' \text{ (see Eq. (2.12))}} \\
&= \left((\underline{\mathbf{a}}'^{(i-1)})^H \underline{u}_i\right)' - \kappa_i (\underline{\mathbf{a}}'^{(i-1)})^H \underline{u}_{i+1} + \kappa_{i-1} \underbrace{(\underline{\mathbf{a}}'^{(i-1)})^H \underline{u}_{i-1}}_{=\kappa_1\kappa_2\cdots\kappa_{i-2}}
\end{aligned} \tag{2.81}$$

However, by using the fact that the vectors $\underline{\mathbf{a}}'^{(i-1)}$ can be expressed as a linear combination of coordinate vectors with no higher order than $(i-1)$, Eq. (2.81) can be rewritten as follows:

$$(2.81) = \left(\underbrace{\left(\sum_{j=1}^{i-1} c_j \underline{u}_j\right)^H \underline{u}_i}_{\text{imaginary}}\right)' - \kappa_i \underbrace{\left(\sum_{j=1}^{i-1} c_j \underline{u}_j\right)^H \underline{u}_{i+1}}_{\text{imaginary}} + \kappa_1\kappa_2\cdots\kappa_{i-2}\kappa_{i-1} \tag{2.82}$$

where c_j are scalar coefficients. Then, by taking the real part of both sides and using the property that

$$\mathrm{Re}\{\underline{u}_j^H \underline{u}_i\} = 0 \quad \text{for } j \neq i \tag{2.83}$$

we have

$$\mathrm{Re}\left\{\left(\underline{\mathbf{a}}'^{(i)}\right)^H \underline{u}_i\right\} = 0 + 0 + \kappa_1\kappa_2\cdots\kappa_{i-2}\kappa_{i-1} \tag{2.84}$$

and this completes the proof. □

2.7.2 *Proof of Theorem 2.1*

Consider a manifold vector described by the following equation:

$$\underline{\mathbf{a}} \triangleq \underline{\mathbf{a}}(p) = \exp(-j(\pi\underline{r}\cos p + \underline{v})) \tag{2.85}$$

The magnitude of the manifold tangent vector is

$$\|\dot{\underline{\mathbf{a}}}(p)\| = \|j\pi \ \sin p \ \underline{r} \odot \underline{\mathbf{a}} \ \| = \pi \sin p \|\underline{r}\| = \dot{s}(p) \tag{2.86}$$

Hence the length of the manifold is

$$l_{\mathrm{m}} \triangleq s(\pi) = \int_0^{\pi} \|\dot{\underline{\mathbf{a}}}(p)\| \, dp = 2\pi \, \|\underline{r}\| \tag{2.87}$$

The first coordinate vector is merely the tangent given by

$$\begin{aligned}\underline{u}_1(s) = \underline{\mathbf{a}}'(s) &= \frac{d\underline{\mathbf{a}}}{ds} = \frac{d\underline{\mathbf{a}}}{dp}\frac{dp}{ds} \\ &= \underbrace{(j\pi \sin p\underline{r} \odot \underline{\mathbf{a}})}_{=\dot{\underline{\mathbf{a}}}(p)} \cdot \underbrace{(\pi \sin p \, \|\underline{r}\|)}_{=\dot{s}(p)}^{-1} \\ &= j\underline{\tilde{r}} \odot \underline{\mathbf{a}}\end{aligned} \tag{2.88}$$

It can be seen that, since the origin of the coordinates is at the array centroid,

$$\underline{\mathbf{a}}^H \underline{u}_1(s) = 0 \tag{2.89}$$

However, the derivative of $\underline{u}_1(s)$ with respect to arc length s is

$$\begin{aligned}\underline{u}_1'(s) &= \frac{d\underline{u}_1}{dp}\frac{dp}{ds} \\ &= \underbrace{(j\underline{\tilde{r}} \odot \dot{\underline{\mathbf{a}}})}_{=\dot{\underline{u}}_1} \cdot \underbrace{(\pi \sin p\|\underline{r}\|)}_{=\dot{s}(p)}^{-1}\end{aligned}$$

$$= j\underline{\tilde{r}} \odot \underbrace{(j\pi \sin p\underline{r} \odot \underline{\mathbf{a}})}_{\dot{\underline{\mathbf{a}}}} \cdot \frac{1}{\pi \sin p\|\underline{r}\|}$$

$$= -\underline{\tilde{r}}^2 \odot \underline{\mathbf{a}} \tag{2.90}$$

Therefore, by using Table 2.2, the first curvature is estimated as

$$\kappa_1 = \|\underline{u}_1'(s)\| = \|\underline{\tilde{r}}^2\| \tag{2.91}$$

Next, the second coordinate vector is chosen to be

$$\underline{u}_2(s) = \frac{\underline{u}_1'(s)}{\|\underline{u}_1'(s)\|}$$

$$= -\frac{1}{\kappa_1}\underline{\tilde{r}}^2 \odot \underline{\mathbf{a}} \tag{2.92}$$

Furthermore,

$$\underline{u}_2'(s) = \frac{d\underline{u}_2}{dp}\frac{dp}{ds}$$

$$= \underbrace{\left(-\frac{1}{\kappa_1}\underline{\tilde{r}}^2 \odot \dot{\underline{\mathbf{a}}}\right)}_{=\dot{\underline{u}}_2} \underbrace{(\pi \sin p\,\|\underline{r}\|)^{-1}}_{=\dot{s}(p)}$$

$$= -\frac{1}{\kappa_1}\underline{\tilde{r}}^2 \odot \underbrace{(j\pi \sin p\underline{r} \odot \underline{\mathbf{a}})}_{\dot{\underline{\mathbf{a}}}} \cdot \frac{1}{\pi \sin p\|\underline{r}\|}$$

$$= -j\frac{1}{\kappa_1}\underline{\tilde{r}}^3 \odot \underline{\mathbf{a}} \tag{2.93}$$

and

$$\underline{u}_2'(s) + \kappa_1\underline{u}_1(s) = -j\left(\frac{1}{\kappa_1}\underline{\tilde{r}}^3 - \kappa_1\underline{\tilde{r}}\right) \odot \underline{\mathbf{a}} \tag{2.94}$$

Hence, the second curvature (see Table 2.2) and the third coordinate vector are

$$\kappa_2 = \|\underline{u}_2'(s) + \kappa_1\underline{u}_1(s)\|$$

$$= \frac{1}{\kappa_1} \cdot \|\,\underline{\tilde{r}}^3 - \kappa_1^2\,\underline{\tilde{r}}\| \tag{2.95}$$

and

$$\begin{aligned}\underline{u}_3(s) &= \frac{1}{\kappa_2}(\underline{u}_2'(s) + \kappa_1 \underline{u}_1(s)) \\ &= -j\frac{1}{\kappa_2} \cdot \left(\frac{1}{\kappa_1}\underline{\tilde{r}}^3 - \kappa_1 \underline{\tilde{r}}\right) \odot \underline{\mathbf{a}} \\ &= -j\frac{1}{\kappa_1\kappa_2} \cdot (\underline{\tilde{r}}^3 - \kappa_1^2 \underline{\tilde{r}}) \odot \underline{\mathbf{a}} \end{aligned} \tag{2.96}$$

In order to estimate the third curvature it is necessary to estimate the terms $\underline{u}_3'(s)$ and $\kappa_2 \underline{u}_2(s)$. Indeed,

$$\begin{aligned}\underline{u}_3'(s) &= \frac{d\underline{u}_3}{dp}\frac{dp}{ds} \\ &= \underbrace{-j\frac{1}{\kappa_1\kappa_2} \cdot (\underline{\tilde{r}}^3 - \kappa_1^2\underline{\tilde{r}}) \odot \underline{\dot{\mathbf{a}}}}_{=\underline{\dot{u}}_3} \cdot \underbrace{(\pi \sin p \|\underline{r}\|)}_{=\dot{s}(p)}{}^{-1} \\ &= -j\frac{1}{\kappa_1\kappa_2} \cdot (\underline{\tilde{r}}^3 - \kappa_1^2\underline{\tilde{r}}) \odot \underbrace{(j\pi \sin p \underline{r} \odot \underline{\mathbf{a}})}_{=\underline{\dot{\mathbf{a}}}}(\pi \sin p \|\underline{r}\|)^{-1} \\ &= \frac{1}{\kappa_1\kappa_2} \cdot (\underline{\tilde{r}}^4 - \kappa_1^2 \underline{\tilde{r}}^2) \odot \underline{\mathbf{a}} \end{aligned} \tag{2.97}$$

and

$$\kappa_2 \underline{u}_2(s) = -\frac{\kappa_2}{\kappa_1}\underline{\tilde{r}}^2 \odot \underline{\mathbf{a}} \tag{2.98}$$

which implies

$$\underline{u}_3'(s) + \kappa_2 \underline{u}_2(s) = \frac{1}{\kappa_1\kappa_2}\left(\underline{\tilde{r}}^4 - (\kappa_1^2 + \kappa_2^2)\underline{\tilde{r}}^2\right) \odot \underline{\mathbf{a}} \tag{2.99}$$

By taking the magnitude of the above equation the third curvature is estimated as

$$\begin{aligned}\kappa_3 &= \|\underline{u}_3'(s) + \kappa_2 \underline{u}_2(s)\| \\ &= \frac{1}{\kappa_1\kappa_2}\|\underline{\tilde{r}}^4 - (\kappa_1^2 + \kappa_2^2)\underline{\tilde{r}}^2\| \end{aligned} \tag{2.100}$$

and, thus, the fourth coordinate vector is

$$\begin{aligned}\underline{u}_4(s) &= \frac{1}{\kappa_3}(\underline{u}_3'(s) + \kappa_2 \underline{u}_2(s)) \\ &= \frac{1}{\kappa_1\kappa_2\kappa_3}\left(\underline{\tilde{r}}^4 - (\kappa_1^2 + \kappa_2^2)\underline{\tilde{r}}^2\right) \odot \underline{\mathbf{a}} \end{aligned} \tag{2.101}$$

By continuing in the same way as above, it can be shown that

(i) the fourth curvature, for $N > 4$, is

$$\kappa_4 = \|\underline{u}_4'(s) + \kappa_3 \underline{u}_3(s)\| = \frac{1}{\kappa_1\kappa_2\kappa_3} \left\| \begin{matrix} \underline{\tilde{r}}^5 \\ -(\kappa_1^2+\kappa_2^2+\kappa_3^2)\underline{\tilde{r}}^3 \\ +(\kappa_1\kappa_3)^2\,\underline{\tilde{r}} \end{matrix} \right\| \tag{2.102}$$

so that the fifth coordinate vector is

$$\underline{u}_5(s) = \frac{1}{\kappa_4}(\underline{u}_4'(s) + \kappa_3\underline{u}_3(s)) = j\frac{1}{\kappa_1\kappa_2\kappa_3\kappa_4} \begin{pmatrix} \underline{\tilde{r}}^5 \\ -(\kappa_1^2+\kappa_2^2+\kappa_3^2)\underline{\tilde{r}}^3 \\ +(\kappa_1\kappa_3)^2\,\underline{\tilde{r}} \end{pmatrix} \odot \underline{\mathbf{a}} \tag{2.103}$$

(ii) the fifth curvature, for $N > 5$, is

$$\kappa_5 = \|\underline{u}_5'(s) + \kappa_4\underline{u}_4(s)\| = \frac{1}{\kappa_1\kappa_2\kappa_3\kappa_4} \left\| \begin{matrix} \underline{\tilde{r}}^6 \\ -(\kappa_1^2+\kappa_2^2+\kappa_3^2+\kappa_4^2)\underline{\tilde{r}}^4 \\ + \begin{pmatrix} (\kappa_1\kappa_3)^2 + (\kappa_1\kappa_4)^2 \\ +(\kappa_2\kappa_4)^2 \end{pmatrix} \underline{\tilde{r}}^2 \end{matrix} \right\| \tag{2.104}$$

so that the sixth coordinate vector is

$$\underline{u}_6(s) = \frac{1}{\kappa_5}(\underline{u}_5'(s) + \kappa_4\underline{u}_4(s)) = -\frac{1}{\kappa_1\kappa_2\kappa_3\kappa_4\kappa_5} \begin{pmatrix} \underline{\tilde{r}}^6 \\ -(\kappa_1^2+\kappa_2^2+\kappa_3^2+\kappa_4^2)\underline{\tilde{r}}^4 \\ + \begin{pmatrix} (\kappa_1\kappa_3)^2 + (\kappa_1\kappa_4)^2 \\ +(\kappa_2\kappa_4)^2 \end{pmatrix} \underline{\tilde{r}}^2 \end{pmatrix} \odot \underline{\mathbf{a}} \tag{2.105}$$

The above equations can be generalized for any index i, to give the curvature κ_i and the vectors $\underline{u}_i$ by the following expressions:

$$\kappa_i = \frac{1}{\kappa_1\kappa_2\cdots\kappa_{i-1}} \left\| \underline{\tilde{r}}^{i+1} - \underline{\tilde{r}}^{i-1} \cdot \sum_{n=1}^{i-1}\kappa_n^2 + \underline{\tilde{r}}^{i-3} \cdot \sum_{n=1}^{i-3}\sum_{\ell=n+2}^{i-1}(\kappa_n\kappa_\ell)^2 - \cdots \right\| \tag{2.106}$$

or, equivalently,

$$\kappa_i = \frac{1}{\kappa_1 \kappa_2 \cdots \kappa_{i-1}} \left\| \sum_{n=1}^{\mathrm{fix}(\frac{i}{2})+1} (-1)^{n-1} \, b_{i,n} \; \underline{\widetilde{r}}^{\,i-2n+3} \right\| \tag{2.107}$$

and

$$\underline{u}_i(s) = \frac{(j)^i}{\kappa_1 \kappa_2 \cdots \kappa_{i-1}} \sum_{n=1}^{\mathrm{fix}(\frac{i-1}{2})+1} (-1)^{n-1} \, b_{i-1,n} \; \underline{\widetilde{r}}^{\,i-2n+2} \odot \underline{\mathbf{a}}(s) \tag{2.108}$$

where $b_{i,n}$ is given by Eq. (2.40).

From Eq. (2.106), it can be seen that the terms inside the norm operator form a polynomial of the sensor locations with coefficients given by the curvatures. This polynomial is expressed in a more compact form in Eq. (2.39). Furthermore, note that the numbers $b_{i,n}$ satisfying the recursive Eq. (2.41) can be easily proven by induction. □

Chapter 3

Differential Geometry of Array Manifold Surfaces

The locus of all array manifold vectors $\underline{\mathbf{a}}(p,q)\ \forall(p,q)$ forms a surface known as the array manifold. This is a two-parameter manifold embedded in an N-dimensional complex space and is formally defined as

$$\mathcal{M} = \{\underline{\mathbf{a}}(p,q) \in \mathfrak{C}^N,\ \ \forall(p,q) : p,q \in \Omega\} \tag{3.1}$$

where Ω denotes the parameter space. The shape of the surface is very important. In the same way that the shape of a curve $\mathcal{A}$ is uniquely defined by its curvatures, the shape of a two-parameter manifold (surface) may be quantitatively expressed in terms of intrinsic geometrical parameters such as the Gaussian curvature $K_G(p,q)$ of the surface (see Fig. 3.1) and the geodesic curvature $\kappa_g(p,q)$ of the curves lying on the surface (see Fig. 3.2).

More specifically, for a point $\underline{\mathbf{a}} \triangleq \underline{\mathbf{a}}(p,q)$ on the manifold surface, the parameters of interest, intrinsic to the manifold surface, together with necessary building blocks of surface differential geometry, will be identified in this chapter and extended from three-dimensional real space to N-dimensional complex space. In order to achieve this all the ramifications and subtleties involved in this process have been taken into consideration.

Thus, it is first necessary to introduce the building blocks of surface differential geometry such as the *manifold metric, first fundamental coefficients* and the *Christoffel symbols.*

However, to proceed, it is essential to start with two definitions which will aid in the development of the basic theory. These are

- the definition of a *constant parameter curve on* $\mathcal{M}$,
- the "regularity" condition by defining the array manifold (surface) as a "vector" function.

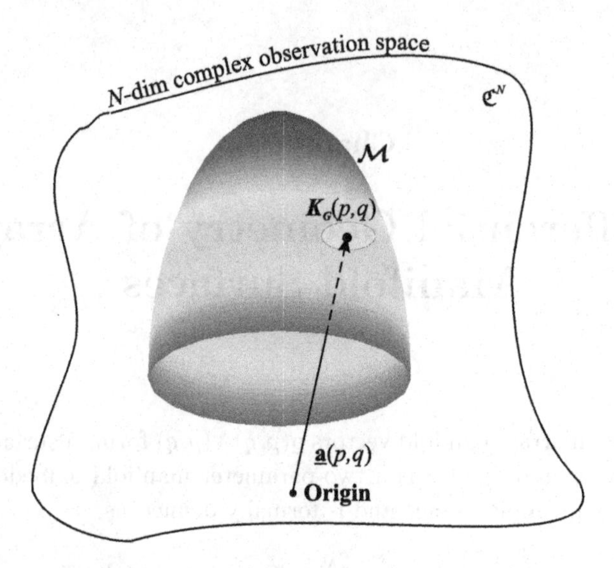

Fig. 3.1 The Gaussian curvature $K_G(p, q)$ provides information about the local shape of the surface $\mathcal{M}$ in the neighbourhood of $\underline{\mathbf{a}}(p, q)$.

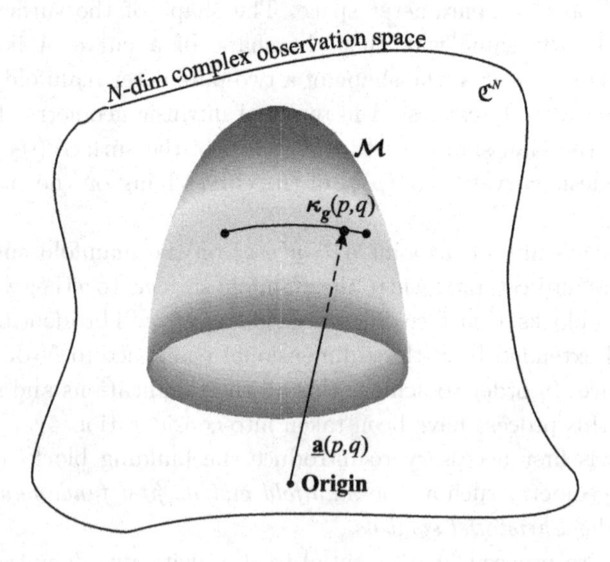

Fig. 3.2 The geodesic curvature κ_g provides information about a curve on a surface.

Definition 3.1 A constant-parameter curve is defined as the curve that joins all those points on the manifold surface $\mathcal{M}$ corresponding to a constant value of one of the two parameters p, q.

Definition 3.2 A regular parametric representation of a surface $\mathcal{M}$ is defined as a vector function

$$\underline{\mathbf{a}} = \underline{\mathbf{a}}(p, q) \tag{3.2}$$

of parameters p and q in the parameter space Ω, if and only if $\forall p, q \in \Omega$, the derivatives $\dot{\underline{\mathbf{a}}}_p = d\underline{\mathbf{a}}/dp$ and $\dot{\underline{\mathbf{a}}}_q = d\underline{\mathbf{a}}/dq$ exist, are continuous and non-zero with

$$\mathbf{rank}\{[\dot{\underline{\mathbf{a}}}_p, \dot{\underline{\mathbf{a}}}_q]\} = 2 \tag{3.3}$$

The vectors $\dot{\underline{\mathbf{a}}}_p$ and $\dot{\underline{\mathbf{a}}}_q$ at a specific point A on the manifold represent the tangent vectors to the p- and q-parameter curves respectively passing through A and also form a basis for the tangent plane to the surface at that point, i.e.

$$\text{Tangent plane} = \mathcal{L}\left[\mathbb{T}\right] \quad \text{where} \quad \mathbb{T} = [\dot{\underline{\mathbf{a}}}_p, \dot{\underline{\mathbf{a}}}_q] \in \mathfrak{C}^{N\times 2} \tag{3.4}$$

The regularity condition of Eq. (3.3) ensures that a tangent plane exists at all points on the surface. Note that the basis created by $\mathbb{T}$ is, in general, not orthonormal.

3.1 Manifold Metric

Let $\underline{\mathbf{a}} = \underline{\mathbf{a}}(p, q)$ be the regular parametric representation of a surface $\mathcal{M}$ embedded in an N-dimensional complex space.

Then the (2×2) real semipositive definite symmetric matrix $\mathbb{G}$, defined as

$$\mathbb{G} \triangleq \operatorname{Re}\left\{\mathbb{T}^H\mathbb{T}\right\} = \begin{bmatrix} \|\dot{\underline{\mathbf{a}}}_p\|^2, & \operatorname{Re}\left\{\dot{\underline{\mathbf{a}}}_p^H \dot{\underline{\mathbf{a}}}_q\right\} \\ \operatorname{Re}\left\{\dot{\underline{\mathbf{a}}}_q^H \dot{\underline{\mathbf{a}}}_p\right\}, & \|\dot{\underline{\mathbf{a}}}_q\|^2 \end{bmatrix} = \begin{bmatrix} g_{pp}, & g_{pq} \\ g_{qp}, & g_{qq} \end{bmatrix} \tag{3.5}$$

is said to be the *manifold metric*. The elements g_{pp}, g_{qq}, and g_{pq} (note $g_{qp} = g_{pq}$) of $\mathbb{G}$ are known, in Differential Geometry terms, as the first fundamental coefficients (or metric coefficients) expressing the magnitudes and inner products of the parameter-curve tangent vectors and entirely describing the manifold properties of the surface. For instance, the angle between the two parameter curves of the surface at a point (p, q) can be expressed as

$$\measuredangle\left(\dot{\underline{\mathbf{a}}}_p, \dot{\underline{\mathbf{a}}}_q\right) = \arccos\left(\frac{g_{pq}}{\sqrt{g_{pp}\, g_{qq}}}\right) \tag{3.6}$$

The metric coefficients provide a way of measuring trajectories on non-Euclidean spaces through the use of weighting coefficients. For example, an infinitesimal distance on the Euclidean plane is measured by

$$ds^2 = dx^2 + dy^2 \tag{3.7}$$

whereas on a general non-Euclidean surface it is given by

$$ds^2 = g_{pp}dp^2 + 2g_{pq}dpdq + g_{qq}dq^2 \tag{3.8}$$

It is also customary to use superscripts to denote the elements of the inverse matrix $\mathbb{G}^{-1}$

$$\mathbb{G}^{-1} = \begin{bmatrix} g^{pp} & g^{pq} \\ g^{qp} & g^{qq} \end{bmatrix} \tag{3.9}$$

where clearly

$$g^{pp} = \frac{g_{qq}}{\det(\mathbb{G})}, \qquad g^{qq} = \frac{g_{pp}}{\det(\mathbb{G})}, \qquad g^{pq} = g^{qp} = \frac{-g_{pq}}{\det(\mathbb{G})} \tag{3.10}$$

while the following properties may be easily proven:

(i) $g_{pp}, g_{qq} \geq 0$
(ii) $\det(\mathbb{G}) \geq 0$ (by Schwarz's inequality) (3.11)
(iii) g_{pp}, g_{pq} and g_{qq} are variant under parameter transformation

---

Note that the metric $\mathbb{G}$ is, by construction, a semipositive definite matrix, i.e.

$$\underline{x}^T \mathbb{G} \underline{x} \geqslant 0, \quad \forall \underline{x} \in \mathcal{R}^2 \tag{3.12}$$

which implies that at some points on the surface $\mathcal{M}$ this matrix can possibly be singular in order to account for the fact that there may not exist a unique tangent space ($\mathcal{L}[\mathbb{T}]$) at each point. That is, at a point (p, q) with a non-unique tangent space, the matrix $\mathbb{G}$ is singular. These points are the boundaries of the manifold surface.

3.2 The First Fundamental Form

Again, let $\underline{a} = \underline{a}(p, q)$ be the regular parametric representation of a manifold surface $\mathcal{M}$ embedded in N-dimensional complex space. Then the differential mapping $d\underline{a} = \dot{\underline{a}}_p dp + \dot{\underline{a}}_q dq$ maps increments dp and dq on the (p, q) parameter plane on to vector $\dot{\underline{a}}_p dp + \dot{\underline{a}}_q dq$ on the tangent plane at $\underline{a}(p, q)$.

The scalar $\mathcal{I}$, defined as

$$\mathcal{I} = \|d\underline{\mathbf{a}}\|^2 = \left\|\dot{\underline{\mathbf{a}}}_p dp + \dot{\underline{\mathbf{a}}}_q dq\right\|^2 \tag{3.13}$$

$$= \begin{bmatrix} dp\ dq \end{bmatrix} \mathbb{G} \begin{bmatrix} dp \\ dq \end{bmatrix} = d\underline{\mathbf{p}}^T \mathbb{G} d\underline{\mathbf{p}} \tag{3.14}$$

with

$$d\underline{\mathbf{p}} = \begin{bmatrix} dp,\ dq \end{bmatrix}^T$$

is known as the first fundamental form and essentially represents the distance between two neighboring points $\underline{\mathbf{a}}(p,q)$ and $\underline{\mathbf{a}}(p+dp, q+dq)$ on the manifold surface.

Two properties of $\mathcal{I}$ which may be readily established are

(i) $\mathcal{I} \geq 0$ with equality when $dp = dq = 0$
(ii) $\mathcal{I}$ is invariant under a parameter transformation (3.15)

The first fundamental form $\mathcal{I}$ is essential for the evaluation of lengths of curves, and areas, on the manifold surface. For instance, let $\underline{\mathbf{a}}(t) \triangleq \underline{\mathbf{a}}(p(t), q(t))$ with $t_1 \leq t \leq t_2$ be a regular curve lying on a surface $\underline{\mathbf{a}} = \underline{\mathbf{a}}(p,q)$. Then clearly the length of the curve is given by

$$\text{length of curve} = \int_{t_1}^{t_2} \left\| \frac{d\underline{\mathbf{a}}}{dt} \right\| dt = \int_{t_1}^{t_2} \sqrt{\frac{\mathcal{I}}{dt^2}}\, dt \tag{3.16}$$

It may be similarly shown that the area of a segment W on the surface of a manifold $\underline{\mathbf{a}} = \underline{\mathbf{a}}(p,q)$ is given by

$$\text{area of segment} = \iint_W \sqrt{\det(\mathbb{G})}\, dp\, dq \tag{3.17}$$

The significance of Eqs. (3.16) and (3.17) in relation to the intrinsic geometry of a surface $\mathcal{M}$ will be clarified later on in this chapter. However, it is clear from Eq. (3.17) that $\sqrt{\det(\mathbb{G})}\, dp\, dq$ describes an infinitesimally small area on the manifold surface which is a function of (p,q) and the array geometry $\mathbf{r}$. Small uncertainties, or variations, in the sensor positions will invariably result in small variations in the infinitesimal area $\sqrt{\det(\mathbb{G})}\, dp\, dq$ on the manifold at point (p,q). The parameter $\sqrt{\det(\mathbb{G})}$ can hence serve as a tool for detecting the changing shape of the manifold surface [18].

3.3 Christoffel Symbol Matrices

In order to investigate the differential geometry of a surface $\underline{\mathbf{a}}(p,q)$, it is necessary to determine how the non-orthonormal basis $\mathbb{T}$ of the tangent plane varies from point to point on the array manifold. This information can be expressed in terms of the *Christoffel symbols of first kind.*

By using the matrix $\mathbb{T} = [\underline{\dot{\mathbf{a}}}_p, \underline{\dot{\mathbf{a}}}_q]$ and its two derivatives $\dot{\mathbb{T}}_p$ and $\dot{\mathbb{T}}_q$ with

$$\dot{\mathbb{T}}_\zeta = d\mathbb{T}/d\zeta = [\underline{\ddot{\mathbf{a}}}_{p\zeta}, \underline{\ddot{\mathbf{a}}}_{q\zeta}] \quad \text{with} \quad \zeta = p \text{ or } q \tag{3.18}$$

the *Christoffel symbols of first kind* represent the inner products between the tangent vectors $\underline{\dot{\mathbf{a}}}_p$, $\underline{\dot{\mathbf{a}}}_q$ (columns of $\mathbb{T}$) and their derivatives (columns of $\dot{\mathbb{T}}_p$ and $\dot{\mathbb{T}}_q$), and are defined as

$$\Gamma_{i,jk} \triangleq \mathrm{Re}\left\{\underline{\dot{\mathbf{a}}}_i^H \underline{\ddot{\mathbf{a}}}_{jk}\right\} \quad \text{with} \quad i,j,k = p \text{ or } q \tag{3.19}$$

Thus there are eight *Christoffel symbols of first kind* forming the two *Christoffel matrices of the first kind* $\mathbf{\Gamma}_{1p}$ and $\mathbf{\Gamma}_{1q}$ defined as follows:

$$\mathbf{\Gamma}_{1\zeta} \triangleq \mathrm{Re}\left\{\mathbb{T}^H \dot{\mathbb{T}}_\zeta\right\} = \begin{bmatrix} \Gamma_{p,p\zeta} & \Gamma_{p,q\zeta} \\ \Gamma_{q,p\zeta} & \Gamma_{q,q\zeta} \end{bmatrix} \quad \text{with} \quad \zeta = p \text{ or } q \tag{3.20}$$

where, due to the symmetry inherent in differentiation,

$$\Gamma_{p,qp} = \Gamma_{p,pq} \quad \text{and} \quad \Gamma_{q,qp} = \Gamma_{q,pq}. \tag{3.21}$$

Although the differential geometry of a surface can be completely described in terms of the above symbols, the process may be considerably simplified by the use of two other matrices $\mathbf{\Gamma}_{2p}$ and $\mathbf{\Gamma}_{2q}$ which are related to the symbols of the first kind $\mathbf{\Gamma}_{1p}$ and $\mathbf{\Gamma}_{1q}$, respectively, in the following way:

$$\mathbf{\Gamma}_{2\zeta} \triangleq \mathbb{G}^{-1}\mathbf{\Gamma}_{1\zeta} = \begin{bmatrix} \Gamma^p_{p\zeta} & \Gamma^p_{q\zeta} \\ \Gamma^q_{p\zeta} & \Gamma^q_{q\zeta} \end{bmatrix} \quad \text{with} \quad \zeta = p \text{ or } q \tag{3.22}$$

These matrices are known as the *Christoffel symbol matrices of second kind* and are used as a transformation which provides the infinitesimal variation of the tangent plane as a point moves on the manifold surface, namely

$$d\mathbb{T} = \mathbb{T}(\mathbf{\Gamma}_{2p}dp + \mathbf{\Gamma}_{2q}dq) \tag{3.23}$$

Once again, due to symmetry,

$$\Gamma^p_{pq} = \Gamma^p_{qp} \quad \text{and} \quad \Gamma^q_{pq} = \Gamma^q_{qp} \tag{3.24}$$

It is obvious that all Christoffel symbols are functions only of the first fundamental coefficients g_{ij} $(i,j = p,q)$ and their derivatives. Thus, for this reason they play a central role in expressing the intrinsic geometry of a surface, which is described in the next section.

3.4 Intrinsic Geometry of a Surface

Let us assume that

$$\mathcal{A}_t \triangleq \{\underline{\mathbf{a}}(t) \in \mathfrak{C}^N, \ \ \forall t : t_1 < t < t_2\} \tag{3.25}$$

where

$$\underline{\mathbf{a}}(t) \triangleq \underline{\mathbf{a}}(p(t), q(t)) \tag{3.26}$$

denotes an arbitrary regular curve on a surface $\mathcal{M}$

$$\mathcal{M} = \{\underline{\mathbf{a}}(p,q) \in \mathfrak{C}^N, \ \ \forall (p,q) : p,q \in \Omega\} \tag{3.27}$$

where Ω denotes the parameter space.

In order to define the intrinsic geometry of a surface, it is first necessary to appreciate the concept of an isometric mapping.

Definition 3.3 Isometry: A one-to-one mapping of a surface $\mathcal{M}$ on to another surface $\breve{\mathcal{M}}$ is called an isometric mapping, or isometry, if the length of an arbitrary curve $\mathcal{A}_t$ on $\mathcal{M}$ is equal to the length of its image $\breve{\mathcal{A}}_t$ on $\breve{\mathcal{M}}$,

$$\text{where} \begin{cases} \breve{\mathcal{A}}_t \triangleq \{\breve{\underline{\mathbf{a}}}(t) \in \mathfrak{C}^K, \ \ \forall t : t_1 < t < t_2\} \\ \breve{\mathcal{M}} \triangleq \{\breve{\underline{\mathbf{a}}}(p,q) \in \mathfrak{C}^K, \ \ \forall (p,q) : p,q \in \Omega\} \end{cases}$$

Hence we may think of isometry as the bending of a surface into a different shape without changing the distance (along the surface) between any of its points. Consequently, the "inhabitants" of such a surface would not be aware of any change at all, as their geometric measurements remain exactly the same. The combination of the above definition with Eqs. (3.16) and (3.17) implies that:

Corollary 3.1 *A one-to-one mapping of a surface $\mathcal{M}$ on to another surface $\breve{\mathcal{M}}$ is an isometry if and only if at corresponding points $\underline{\mathbf{a}}(p,q)$ and $\breve{\underline{\mathbf{a}}}(p,q)$ the first fundamental coefficients remain unchanged, i.e. $g_{ij} = \breve{g}_{ij}$ $\forall i,j = p,q$.*

Thus, there is no difference in the measurement of lengths, angles, and areas on isometric surfaces although the surfaces, when viewed from the embedding space, may have entirely different geometric shapes. **An isometric mapping is by necessity both angle-preserving (conformal) and area-preserving (equiareal).**

Definition 3.4 Intrinsic Property and Intrinsic Geometry: A property of a surface which remains invariant under an isometry is called an intrinsic property of the surface. The totality of the intrinsic properties of a surface is known as the intrinsic geometry of the surface.

Naturally a property of a surface is an intrinsic property if it only depends on the first fundamental coefficients (and hence the magnitudes and inner products of the tangents to the surface). *The intrinsic geometry of a surface is completely independent of the space in which the surface is embedded and is built from solely isometric invariants.*

Next, two important features of intrinsic geometry are presented, namely:

- Gaussian curvature K_G, and,
- geodesic curvature κ_g.

3.4.1 *Gaussian Curvature*

By considering the surface as a whole body with intrinsic properties, Gauss, according to his "Theorema Egregium", assigned to every point on the surface $\underline{a}(p,q)$, independently of any specific curve passing through it, a real number K_G called Gaussian curvature. Research into surfaces has produced several formulas for K_G. For example, the Gaussian curvature for surfaces embedded in the 3-dimensional Euclidian space $\mathcal{R}^3$ is given below (see [2]).

$$K_G(p,q) = \frac{\left(\ddot{\underline{a}}_p^T \underline{T}\right) \cdot \left(\ddot{\underline{a}}_q^T \underline{T}\right) - \left(\frac{d^2\underline{a}}{dp\,dq}\right)^T \underline{T}}{\det(\mathbb{G})} \tag{3.28}$$

where $\underline{T} \triangleq \underline{T}(p,q)$ is the unique normal to the surface at point (p,q).

As the array manifold is embedded in $\mathfrak{C}^N$ (and not in R^3) a unique normal does not exist to each point and, therefore, Eq. (3.28) cannot be defined for array manifolds. Here we will only consider the intrinsic formula for the Gaussian curvature which is independent of the normal space. For instance, an intrinsic formula for the Gaussian curvature, $K_G(p,q)$, of a surface, can be shown to be of the form [17] (also known as the *curvature scalar*):

$$K_G(p,q) = \frac{1}{\det(\mathbb{G})}\begin{pmatrix} \frac{1}{2}\left(2\frac{d^2 g_{pq}}{dp\,dq} - \frac{d^2 g_{pp}}{dq^2} - \frac{d^2 g_{qq}}{dp^2}\right) \\ + \sum_{i,j=p,q} g^{ij} \det\begin{bmatrix} \Gamma_{i,pq} & \Gamma_{i,pp} \\ \Gamma_{j,qq} & \Gamma_{j,qp} \end{bmatrix} \end{pmatrix} \tag{3.29}$$

The above expression may be more compactly represented in terms of the Christoffel symbols:

$$K_G(p,q) = \frac{1}{\det(\mathbb{G})}\begin{pmatrix} \frac{d}{dp}\Gamma_{p,qq} - \frac{d}{dq}\Gamma_{p,qp} \\ + \mathrm{col}_2\{\mathbf{\Gamma}_{2p}\}^T \mathrm{col}_1\{\mathbf{\Gamma}_{1q}\} \\ - \mathrm{col}_2\{\mathbf{\Gamma}_{2q}\}^T \mathrm{col}_1\{\mathbf{\Gamma}_{1p}\} \end{pmatrix} \tag{3.30}$$

where $\mathrm{col}_i\{matrix\}$ denotes the ith column of the *matrix*.

Here the following intrinsic formula is used, adapted from [19], and expressed as a function of the metric $\mathbb{G}$ as follows:

$$K_G(p,q) = -\frac{1}{\sqrt{\det(\mathbb{G})}}\left(\frac{d\left(\frac{\sqrt{\det(\mathbb{G})}}{g_{pp}}\Gamma^q_{pq}\right)}{dp} - \frac{d\left(\frac{\sqrt{\det(\mathbb{G})}}{g_{pp}}\Gamma^q_{pp}\right)}{dq}\right) \tag{3.31}$$

where Γ^q_{pq}, Γ^q_{pp} are elements of the 2×2 real matrices $\mathbf{\Gamma}_{2p}$ and $\mathbf{\Gamma}_{2q}$.

The sign of the Gaussian curvature of Eq. (3.31) or (3.29), provides an indication of the local shape of the surface in that neighborhood. For instance, the surface around a point (p,q) is locally:

- elliptic, if $K_G(p,q) > 0$ (the whole neighborhood of the surface at the point considered lies on one side of the tangent hyperplane),
- hyperbolic, if $K_G(p,q) < 0$ (one part of the surface at the point considered lies on one side of the tangent hyperplane and the other part on the other side),
- either parabolic or planar (i.e. flat), if $K_G(p,q) = 0$ (there is a straight line of the surface, lying totally on the tangent space).

Example 3.1 The following lists a number of surfaces in $\mathcal{R}^3$ and their respective Gaussian curvatures:

(a) Sphere of radius R: $K_G = 1/R^2 \Rightarrow$ Surface is elliptic at every point.
(b) Cylinder or Cone: $K_G = 0 \Rightarrow$ Surface is parabolic at every point.
(c) 2-D plane: $K_G = 0 \Rightarrow$ Surface is flat at every point.

If the Gaussian curvature at a point on the manifold is a positive number, the local neighborhood of that point is equivalent to the local neighborhood of a point on a sphere of radius $\sqrt{1/K_G}$. Also note that surfaces which have a *constant Gaussian curvature* at every point (e.g. Parts (b) and (c) of Example 3.1 above — a zero Gaussian curvature at every point) are said to be *developable surfaces* and will be discussed later in Section 3.5.

Apart from defining the local shape of a surface, the Gaussian curvature has important implications with regards to isometric mappings. For example,

two surfaces which are related by an isometric mapping must have the same Gaussian curvature at corresponding points (since K_G is an intrinsic property)

The converse of this is in general not true. However if two surfaces have the same *constant* Gaussian curvature, then any two sufficiently small neighborhoods of the surfaces are related by an isometric mapping. In particular

- if $K_G(p, q) = constant$, $\forall(p, q)$ then, using Minding's theorem [19] which states that two surfaces of the **same constant** curvature are locally isometric, we have the following cases:
  (1) if $K_G = 0$, the surface $\mathcal{M}$ is isometric with **the plane**,
  (2) if $K_G > 0$, the surface $\mathcal{M}$ is isometric with **a sphere** of radius $1/\sqrt{K_G}$, and
  (3) if $K_G < 0$, the surface $\mathcal{M}$ is isometric with **a surface of revolution**, called a pseudo-sphere, determined by the value $1/\sqrt{-K_G}$.

As a result, developable surfaces for which $K_G = 0$ are the only surfaces that can be mapped isometrically *onto a plane*. This feature will be used later in Section 3.5.

3.4.2 *Curves on a Manifold Surface: Geodesic Curvature*

The Gaussian curvature, discussed in the previous section, and the geodesic curvatures are the two most important parameters of the intrinsic geometry of a surface. This means that these two parameters remain invariant under an "isometry" (see Corollary 3.1, page 65).

In this section we will focus our attention on geodesic curvature $\kappa_g(s)$ which is the final differential geometry parameter of interest to be presented in this study, and is associated with curves lying on the manifold surface. However, initially the arc length of a curve on the manifold surface should be defined and then, in order to understand the significance of $\kappa_g(s)$, the geodecity concept should be introduced.

3.4.2.1 *Arc Length*

The arc length s can be interpreted in physical terms as the total distance covered by a person travelling along a certain path on a certain landscape. For a curve $\mathcal{A}_t$

$$\mathcal{A}_t \triangleq \{\underline{\mathbf{a}}(t) \in \mathfrak{C}^N, \ \forall t : t \in \Omega_t\} \tag{3.32}$$

where

$$\underline{\mathbf{a}}(t) \triangleq \underline{\mathbf{a}}(p(t), q(t))$$

on the array manifold surface $\mathcal{M}$, the arc length s is the total distance travelled along the curve $\mathcal{A}_t$ from an "initial" point $(p(t_1), q(t_1))$ to the "current" point $(p(t_2), q(t_2))$ under consideration. Its expression as a function of the manifold metric is, based on Eq. (3.16):

$$\underset{\text{(curve on a surface)}}{\text{arc length}} \quad : \ s = \int_{t_1}^{t_2} \sqrt{\frac{d\underline{\mathrm{p}}}{dt}^T \mathbb{G} \frac{d\underline{\mathrm{p}}}{dt}} dt \tag{3.33}$$

where

$$\underline{\mathrm{p}} = [p(t), q(t)]^T \quad \text{and, hence,} \quad \frac{d\underline{\mathrm{p}}}{dt} = \left[\frac{dp}{dt}, \frac{dq}{dt}\right]^T \tag{3.34}$$

3.4.2.2 *The Concept of Geodicity*

It is well known that the curve with the minimum length between two points in an Euclidian space is a straight line. This concept is extended to a curve connecting two points on a surface and the curve of minimum length belonging to the surface is called a geodesic curve (equivalent to a straight line in a plane). By considering a curve on a surface connecting two points, the closeness of this curve to a geodesic curve can be assessed by means of the geodesic curvature, κ_g, with $\kappa_g = 0$ corresponding to a geodesic curve. That is,

$$\text{geodesic curve} \iff \text{a curve with } \kappa_g = 0 \tag{3.35}$$

3.4.3 *Geodesic Curvature*

We have seen that geodesic curves on an arbitrary geometric surface generalize the notion of a straight line in Euclidean geometry. It is also known that a straight line $\underline{\mathbf{a}}(t) = \underline{x} + t\underline{y}$ is characterized by zero curvature or acceleration $\kappa_1 = \|\underline{\mathbf{a}}''\| = 0$. Following this line of reasoning, the intrinsic acceleration, or geodesic curvature, of a curve is defined as follows:

Definition 3.5 Geodesic Curvature: The geodesic curvature κ_g of a curve $\underline{\mathbf{a}}(t)$ on a surface $\mathcal{M}$ is equal to the component of the curve's first curvature κ_1 along the tangent plane to the surface at every point along the curve.

Clearly the geodesic curvature is a function of the first fundamental coefficients only and is consequently preserved under an isometric mapping. This implies that **an isometric mapping is by necessity also a geodesic mapping.**

Following the above definition, we reach the same conclusion as that of Eq. (3.35), i.e.

> **a geodesic curve on a surface $\mathcal{M}$ may be defined as a curve whose geodesic curvature is zero at every point along its length**

For example, since for a geodesic curve the direction of acceleration is always orthogonal to the surface, the inhabitants of a surface perceive no acceleration at all — for them the geodesic is a "straight" line.

It can also be shown that *in the neighborhood of a point A on a surface, there exists a unique geodesic through A in any given direction. Consequently a unique geodesic joins point A with every point B in its neighborhood.* Furthermore, the geodesic defines a unique arc of minimum length between neighboring points A and B. The converse is also true in that if $\underline{a}(t)$ is an arc of minimum length between any two points on a surface, $\underline{a}(t)$ is a geodesic.

The geodesic curvature can be estimated (for proof, see Appendix 3.7.1) by the following expression:

$$\kappa_g(s) = \sqrt{\det(\mathbb{G})}\; \underline{\mathrm{p}}'^T \begin{bmatrix} 0 & 1 \\ -1 & 0 \end{bmatrix} (\underline{\mathrm{p}}'' + (\mathbf{\Gamma}_{2p} p' + \mathbf{\Gamma}_{2q} q')\, \underline{\mathrm{p}}') \tag{3.36}$$

where

$$\underline{\mathrm{p}} = [p(s), q(s)]^T$$

where s is the arc length of the curve under consideration. From Eq. (3.36), a geodesic curve, i.e. a curve for which the geodesic curvature is invariably zero, must satisfy the equation:

$$\underline{p}'' + (\mathbf{\Gamma}_{2p} p' + \mathbf{\Gamma}_{2q} q')\, \underline{p}' = \underline{0}_2 \quad \forall (p, q) \tag{3.37}$$

Equation (3.37) is a system of two simultaneous equations which are the well-known geodesic differential equations (page 64 [19]).

It is clear that the computation of both the Gaussian and geodesic curvature parameters (see Eqs. (3.31) and (3.36)), involves the elements of the Christoffel Symbol matrices of second kind $\mathbf{\Gamma}_{2p}$ and $\mathbf{\Gamma}_{2q}$.

It is also clear that the general expression for the geodesic curvature κ_g of an arbitrary curve on a surface is rather involved but can be considerably simplified for the special case of constant-parameter curves [2]: Let us denote a "p-parameter curve" (or simply p-curve) on the surface $\mathcal{M}$ corresponding to a constant value of $q = q_o$ as

$$\mathcal{A}_{p|q_o} = \{\underline{\mathbf{a}}(p, q_o) \in \mathfrak{C}^N, \ \forall p : p \in \Omega_p, q_o = \text{constant}\} \tag{3.38}$$

and, in a similar fashion, a "q-parameter curve" (or simply q-curve) as

$$\mathcal{A}_{q|p_o} = \{\underline{\mathbf{a}}(p_o, q) \in \mathfrak{C}^N, \ \forall q : q \in \Omega_q, p_o = \text{constant}\} \tag{3.39}$$

Different values of q_o (or p_o, accordingly) generate a family of curves having the same properties covering the whole of the manifold surface $\mathcal{M}$. Thus, there are two such families of curves providing two alternative ways of treating the manifold surface. These are

- the family of p-parameter curves

$$\mathcal{M} = \{\mathcal{A}_{p|q_o}, \ \forall q_o : q_o \in \Omega_q\} \tag{3.40}$$

- the family of q-parameter curves

$$\mathcal{M} = \{\mathcal{A}_{q|p_o}, \ \forall p_o : p_o \in \Omega_p\} \tag{3.41}$$

Both families of p- and q-curves can be used to describe the manifold surface $\mathcal{M}$ but different parametrizations of the surface provide families of curves with different properties.

Note that if two families of curves are orthogonal and one family consists of geodesic curves, then these two families constitute a set of what is known as "geodesic coordinates."

Based on the above definitions and using Eq. (3.36), the geodesic curvature $\kappa_{g,p}$ for p-curves (i.e. curves of constant parameter $q = q_o$) of an array manifold can be shown, using $dq/ds = 0$ and $dp/ds = 1/\sqrt{g_{pp}}$, to be equal to

$$\kappa_{g,p} \triangleq \kappa_g(p, q_o) = \sqrt{\frac{\det(\mathbb{G})}{g_{pp}^3}}\Gamma_{pp}^{q_o} \tag{3.42}$$

Similarly, for the q-curves (i.e. curves of constant parameter $p = p_o$), using $dp/ds = 0$ and $dq/ds = 1/\sqrt{g_{qq}}$, the geodesic curvature $\kappa_{g,q}$ can be found to be

$$\kappa_{g,q} \triangleq \kappa_g(p_o, q) = -\sqrt{\frac{\det(\mathbb{G})}{g_{qq}^3}}\Gamma_{qq}^{p_o} \tag{3.43}$$

Thus an alternative approach is produced by treating the manifold surface as a family of curves which fully covers and describes the corresponding surface.

In this book (Chapters 4 and 5), two different array manifold parametrizations, having significant differences, will be studied. Each parametrization treats the same array manifold $\mathcal{M}$, using two different families of curves with diverse properties. These are

- the (θ, ϕ) parametrization (or, azimuth-elevation parametrization),
- the (α, β) parametrization (or, cone-angle parametrization).

3.5 The Concept of "Development"

In Section 3.4.1, we have seen that by examining the sign of the Gaussian curvature K_G of the array manifold it is possible to study its shape. It is recognized that apart from the fact that the Gaussian curvature offers an indication about the manifold shape, its importance is enhanced by the fact that if it satisfies certain stringent conditions (valid for a very broad class of arrays) the geodesic mappings of the individual manifold curves result in a consistent mapping of the whole manifold on the parameter plane, called the *development of the manifold* (see Fig. 3.3), which is defined as follows:

Definition 3.6 Development: Let $\mathcal{A}_t$ be a differentiable curve lying on the manifold $\mathcal{M}$ expressed in terms of a parameter t as follows:

$$\mathcal{A}_t = \{\underline{\mathbf{a}}(t) \triangleq \underline{\mathbf{a}}(p(t), q(t)) \in \mathfrak{C}^N, \ \forall \zeta : t \in \Omega_t\} \tag{3.44}$$

Furthermore, let $s(t)$ be the arc length along $\mathcal{A}_t$ given by Eq. (3.33). The *development* of $\mathcal{A}_t$ on the plane $\mathcal{R}^2$ is the plane curve

$$\mathcal{A}_d = \{\underline{\mathbf{a}}_d(t) \in \mathcal{R}^2, \ \forall t : t \in \Omega_\zeta\} \tag{3.45}$$

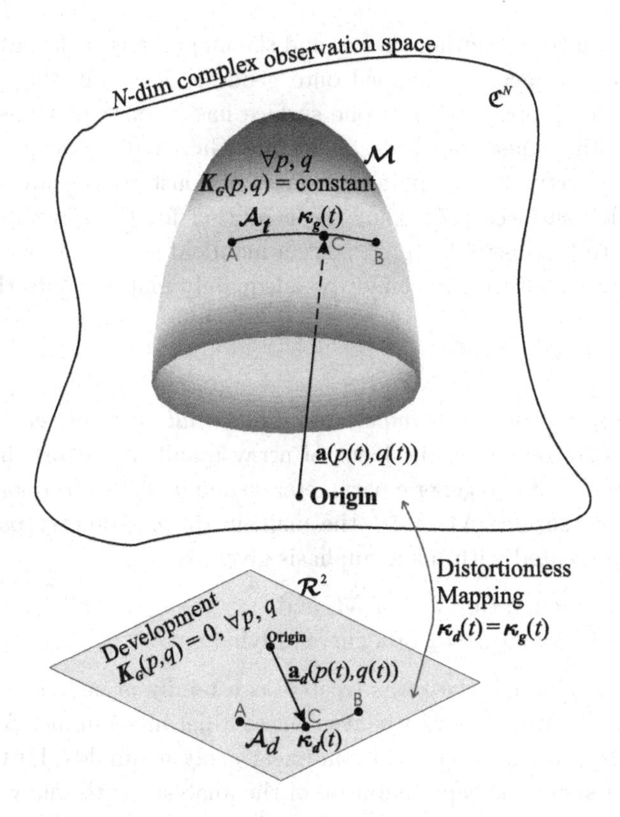

Fig. 3.3 The concept of "development."

with curvature $\kappa_d(t)$ and arc length $s_d(t)$ given by

$$\kappa_d(t) = \kappa_g(t) \quad \text{and} \quad s_d(t) = \int_t \kappa_d^{-1}(t)\, dt \tag{3.46}$$

where

$$\kappa_g(t) \triangleq \kappa_g((p(t), q(t))) = \text{geodesic curvature of } \mathcal{A}_t$$

Thus, **the curvature $\kappa_d(t)$ of the development of a manifold curve A_t is the geodesic curvature of that curve.** The motivation behind the conservation of the geodesic curvatures is to maintain the characteristic that the shortest path between two points on the surface maps to a straight line on the plane (i.e. the shortest path on the development), otherwise the development will represent a "distorted" surface.

The relation between the Gaussian curvature and the "development" is addressed by Beltrami's theorem [20, 21]. This theorem states that if a surface is mapped to another surface and the mapping is such that geodesic curves in one surface are mapped onto geodesic curves in the other surface (geodesic mapping) then if one surface has a constant Gaussian curvature the other must also have a constant Gaussian curvature, in order to have a distortionless mapping. Surfaces of constant K_G are known as "developable" surfaces [17]. Thus the condition for the development[1] to exist is K_G to be constant $\forall(p, q)$, since a manifold surface $\mathcal{M}$ with a varying Gaussian curvature K_G cannot be adequately mapped onto the plane.

3.6 Summary

In this chapter *regularly parametrized differential* surfaces, embedded in $\mathfrak{C}^N$, which may represent the locus of array manifold vectors, have been studied in terms of two generic parameters p and q. Thus for a point $\underline{a}(p, q)$ on a manifold surface $\mathcal{M}$ in $\mathfrak{C}^N$ the main intrinsic geometry parameters have been presented with main emphasis given to

- the Gaussian curvature K_G of $\mathcal{M}$, and,
- the geodesic curvature κ_g of a curve $\mathcal{A}$ lying on $\mathcal{M}$.

Then a manifold surface was treated as a family of curves on the surface $\mathcal{M}$. This treatment is very convenient as it permits a unified framework for the analysis of the linear and non-linear array manifolds. Furthermore, to provide a simplified representation of the analysis, with many potential benefits, the class of developable surfaces has been identified and the concept of the development has been introduced. This concept will be employed in the next chapter to map the array manifold surface, embedded in a complex N-dimensional space, on to a real parameter plane whilst preserving its main intrinsic geometry properties and characteristics. This has potential to simplify the analysis of array manifold surfaces.

3.7 Appendices

3.7.1 *Proof of Eq. (3.36) — Geodesic Curvature*

Consider the locus of the manifold vectors $\underline{a}(p, q)$, $\forall(p, q)$, forming the surface $\mathcal{M}$ embedded in $\mathfrak{C}^N$. Consider a curve $\mathcal{A}$

$$\mathcal{A} \triangleq \{\underline{a}(s) \in \mathfrak{C}^N, \ \forall s : s \in \Omega_s\} \tag{3.47}$$

[1]Note that the "development" is on a plane and planes have constant (equal to zero) Gaussian curvature.

where

$$\underline{\mathbf{a}}(s) \triangleq \underline{\mathbf{a}}(p(s), q(s))$$

lying on $\mathcal{M}$ and let us define $\underline{\mathrm{p}}(s) \triangleq [p(s), q(s)]^T$ where s is the arc length of the curve. Then the tangent vector at a point $(p(s), q(s))$ on the curve $\mathcal{A}$ is given as follows:

$$\underline{\mathbf{a}}' \equiv \frac{d\underline{\mathbf{a}}(s)}{ds} = \mathbb{T}\,\underline{\mathrm{p}}'$$

where $\mathbb{T} = [\underline{\dot{\mathbf{a}}}_p,\ \underline{\dot{\mathbf{a}}}_q] \in \mathfrak{C}^{N\times 2}$ such that $\mathcal{L}[\mathbb{T}]$ is the tangent space to the manifold surface at $(p(s), q(s))$. Furthermore, let us define the matrix $\mathbb{B} = [\underline{b}_1,\ \underline{b}_2] \in \mathfrak{C}^{N\times 2}$ as an orthonormal basis for the space $\mathcal{L}[\mathbb{T}]$ with $\underline{b}_1$ chosen to be equal to $\underline{\mathbf{a}}'/\|\underline{\mathbf{a}}'\|$. This gives the expression:

$$\mathbb{B} = [\underline{b}_1,\ \underline{b}_2] = \mathbb{T}\mathbb{E}_p\mathbb{H} \tag{3.48}$$

where

$$\mathbb{E}_p = \frac{1}{\sqrt{\det(\mathbb{G})}} \begin{bmatrix} \sqrt{\dfrac{\det(\mathbb{G})}{g_{pp}}} & -\dfrac{g_{pq}}{\sqrt{g_{pp}}} \\ 0 & \sqrt{g_{pp}} \end{bmatrix} \in \mathcal{R}^{2\times 2} \tag{3.49}$$

is the Gram-Schmidt orthonormalization matrix, such that the first column of $\mathbb{T}\mathbb{E}_p$ is the basis vector $\underline{\dot{\mathbf{a}}}_p/\|\underline{\dot{\mathbf{a}}}_p\|$ and the matrix $\mathbb{H} \in \mathcal{R}^{2\times 2}$ is the orthogonal rotation matrix:

$$\mathbb{H} = \begin{bmatrix} \cos\psi & -\sin\psi \\ \sin\psi & \cos\psi \end{bmatrix} \tag{3.50}$$

where ψ is the rotation angle between the tangent vector to the p-curve $\underline{\mathbf{a}}_p$ and the tangent vector to the curve under consideration $\underline{\mathbf{a}}'$:

$$\psi = \measuredangle\left(\underline{\dot{\mathbf{a}}}_p, \underline{\mathbf{a}}'\right) = \arccos\left(\frac{\mathrm{Re}\{\underline{\dot{\mathbf{a}}}_p^H\,\underline{\mathbf{a}}'\}}{\|\underline{\dot{\mathbf{a}}}_p\|\ \|\underline{\mathbf{a}}'\|}\right) \tag{3.51}$$

By projecting the second derivative

$$\underline{\mathbf{a}}'' \triangleq \frac{d\,(\underline{\mathbf{a}}')}{ds} = \frac{d\mathbb{T}}{ds}\frac{d\underline{\mathrm{p}}}{ds} + \mathbb{T}\frac{d^2\underline{\mathrm{p}}}{ds^2} \tag{3.52}$$

of the manifold vector at $(p(s), q(s))$ onto the subspace $\mathcal{L}[\mathbb{B}]$ (note that $\mathcal{L}[\mathbb{B}] = \mathcal{L}[\mathbb{T}]$), we have

$$\mathbb{P}_{\mathbb{B}}\underline{\mathbf{a}}'' = \kappa_1\underline{b}_1 + \kappa_g\underline{b}_2 \tag{3.53}$$

where $\mathbb{P}_{\mathbb{B}} = \mathbb{B}\left(\mathbb{B}^H\mathbb{B}\right)^{-1}\mathbb{B}^H$ is the projection operator. The coefficient of $\underline{b}_2$, that is κ_g, is defined as the geodesic curvature of the curve at $(p(s), q(s))$ and represents the component of $\underline{\mathbf{a}}''$ in the direction of $\underline{b}_2$. Hence its expression is

$$\kappa_g = \mathrm{Re}\{\underline{b}_2^H \underline{\mathbf{a}}''\} \tag{3.54}$$

The vector $\underline{b} = [0,\ 1]^T$ will be used for notational purposes such that $\underline{b}_2 = \mathbb{B}\underline{b}$. Substituting the expressions of $\underline{b}_2$ and $\underline{\mathbf{a}}''$ into Eq. (3.54) gives, after some manipulation:

$$\kappa_g(s) = \underline{b}^T \mathbb{H}^T \mathbb{E}_p^T \mathbb{G} \left\{ \left(\mathbf{\Gamma}_{2p} \frac{dp}{ds} + \mathbf{\Gamma}_{2q} \frac{dq}{ds} \right) \frac{d\underline{\mathrm{p}}}{ds} + \frac{d^2\underline{\mathrm{p}}}{ds^2} \right\}. \tag{3.55}$$

In Eq. (3.55) the term $\underline{b}^T \mathbb{H}^T \mathbb{E}_p^T \mathbb{G}$ can be simplified to $\sqrt{\det(\mathbb{G})} \cdot [-p', q']$, providing

$$\kappa_g(s) = \sqrt{\det(\mathbb{G})} \begin{bmatrix} -p' \\ q' \end{bmatrix}^T \left((\mathbf{\Gamma}_{2p} p' + \mathbf{\Gamma}_{2q} q')\, \underline{\mathrm{p}}' + \underline{\mathrm{p}}'' \right) \tag{3.56}$$

□

Chapter 4

Non-Linear Arrays: (θ, ϕ)-Parametrization of Array Manifold Surfaces

It is common practice to express the direction of array signals in terms of azimuth angle θ and elevation angle ϕ. In this case the response vector of an array of N omnidirectional sensors is (see Eq. (1.22))

$$\underline{\mathbf{a}}(\theta, \phi) = \exp\left(-j\mathbf{r}^T \underline{\mathrm{k}}(\theta, \phi)\right) \tag{4.1}$$

where $\mathbf{r}$ is the matrix with columns the sensor location and $\underline{\mathrm{k}}(\theta, \phi)$ is the wavenumber vector given by Eq. (1.3). Thus by using the directional parameters (θ, ϕ), instead of the generic parameters (p, q) of the previous chapter, the two-parameter array manifold $\mathcal{M}$ of Eq. (3.1) becomes

$$\mathcal{M} = \{\underline{\mathbf{a}}(\theta, \phi) \in \mathfrak{C}^N, \ \forall(\theta, \phi) : (\theta, \phi) \in \Omega\} \tag{4.2}$$

where

$$\Omega = \{(\theta, \phi) : \theta \in [0°, 360°) \text{ and } \phi \in [0°, 90°)\}$$

The aim of this chapter is to demonstrate the feasibility and potential benefits of the theoretical framework presented in Chapters 2 and 3 by examining various non-linear arrays of omnidirectional sensors. However, in order to proceed, the following derivatives of the wavenumber vector $\underline{\mathrm{k}}(\theta, \phi)$ are required

$$\text{first derivatives:} \begin{cases} \dot{\underline{\mathrm{k}}}_\theta = \pi[-\sin\theta\cos\phi, \cos\theta\cos\phi, 0]^T \\ \dot{\underline{\mathrm{k}}}_\phi = \pi[-\cos\theta\sin\phi, -\sin\theta\sin\phi, \cos\phi]^T \end{cases} \tag{4.3}$$

$$\text{second derivatives:} \begin{cases} \ddot{\underline{\mathrm{k}}}_{\theta\theta} = \pi[-\cos\theta\cos\phi, -\sin\theta\cos\phi, 0]^T \\ \ddot{\underline{\mathrm{k}}}_{\theta\phi} = -\dot{\underline{\mathrm{k}}}_\theta \tan\phi \\ \ddot{\underline{\mathrm{k}}}_{\phi\phi} = -\underline{\mathrm{k}}(\theta, \phi) \end{cases} \tag{4.4}$$

where the following notation has been used

$$\dot{\underline{k}}_{\mathfrak{p}} = \frac{d\underline{k}(\theta,\phi)}{d\mathfrak{p}} \quad \text{and} \quad \ddot{\underline{k}}_{\mathfrak{pq}} = \frac{d^2\underline{k}(\theta,\phi)}{d\mathfrak{p}\, d\mathfrak{q}} \tag{4.5}$$

with

$$(\mathfrak{p} \text{ and } \mathfrak{q}) = (\theta \text{ or } \phi)$$

4.1 Manifold Metric and Christoffel Symbols

For an azimuth-elevation array system of omnidirectional sensors the function $\underline{a} = \underline{a}(\theta,\phi), \forall(\theta,\phi)$, provides a *regular parametric* representation of the manifold surface $\mathcal{M}$. With $\mathcal{M}$ embedded in an N-dimensional complex space, the manifold metric $\mathbb{G}$ is a function of the array geometry $\mathbf{r}$ and given by the following expression

$$\begin{aligned}\mathbb{G} &= \begin{bmatrix} g_{\theta\theta}, & g_{\phi\theta} \\ g_{\theta\phi}, & g_{\phi\phi} \end{bmatrix} \\ &= [\dot{\underline{k}}_\theta \quad \dot{\underline{k}}_\phi]^T \mathbf{r}\mathbf{r}^T [\dot{\underline{k}}_\theta \quad \dot{\underline{k}}_\phi] \end{aligned} \tag{4.6}$$

Thus, for arrays of omnidirectional sensors, the elements of the metric $\mathbb{G}$ can be computed as

$$g_{\mathfrak{pq}} = \dot{\underline{k}}_{\mathfrak{p}}^T \mathbf{r}\mathbf{r}^T \dot{\underline{k}}_{\mathfrak{q}} \quad \text{where} \quad \mathfrak{p} \text{ and } \mathfrak{q} = \theta \text{ or } \phi \tag{4.7}$$

while the Christoffel symbol matrices are

$$\text{First kind:} \begin{cases} \mathbf{\Gamma}_{1\theta} = \begin{bmatrix} \overbrace{\dot{\underline{k}}_\theta^T \mathbf{r}\mathbf{r}^T \ddot{\underline{k}}_{\theta\theta}}^{=\Gamma_{\theta,\theta\theta}}, & \overbrace{-g_{\theta\theta}\tan\phi}^{=\Gamma_{\theta,\phi\theta}} \\ \underbrace{\dot{\underline{k}}_\phi^T \mathbf{r}\mathbf{r}^T \ddot{\underline{k}}_{\theta\theta}}_{=\Gamma_{\phi,\theta\theta}}, & \underbrace{-g_{\theta\phi}\tan\phi}_{=\Gamma_{\phi,\phi\theta}} \end{bmatrix} \\ \text{and} \\ \mathbf{\Gamma}_{1\phi} = \begin{bmatrix} \overbrace{-g_{\theta\theta}\tan\phi}^{=\Gamma_{\theta,\theta\phi}}, & \overbrace{-\underline{k}^T \mathbf{r}\mathbf{r}^T \dot{\underline{k}}_\theta}^{=\Gamma_{\theta,\phi\phi}} \\ \underbrace{-g_{\theta\phi}\tan\phi}_{=\Gamma_{\phi,\theta\phi}}, & \underbrace{-\underline{k}^T \mathbf{r}\mathbf{r}^T \dot{\underline{k}}_\phi}_{=\Gamma_{\phi,\phi\phi}} \end{bmatrix} \end{cases} \tag{4.8}$$

$$\text{Second kind:} \begin{cases} \boldsymbol{\Gamma}_{2\theta} = \begin{bmatrix} \Gamma^{\theta}_{\theta\theta}, & \overbrace{-\tan\phi}^{=\Gamma^{\theta}_{\phi\theta}} \\ \Gamma^{\phi}_{\theta\theta}, & \underbrace{0}_{=\Gamma^{\phi}_{\phi\theta}} \end{bmatrix} \\ \text{and} \\ \boldsymbol{\Gamma}_{2\phi} = \begin{bmatrix} \overbrace{-\tan\phi}^{=\Gamma^{\theta}_{\theta\phi}}, & \Gamma^{\theta}_{\phi\phi} \\ \underbrace{0}_{=\Gamma^{\phi}_{\theta\phi}}, & \Gamma^{\phi}_{\phi\phi} \end{bmatrix} \end{cases} \tag{4.9}$$

It should be noted from Eq. (4.9) that $\Gamma^{\phi}_{\theta\phi} = 0$. This simplifies the Gaussian curvature of Eq. (3.31), for a 3D array of N-omnidirectional sensors, to

$$K_G(\theta, \phi) = \frac{1}{\sqrt{\det(\mathbb{G})}} \frac{d}{d\phi} \left(\frac{\sqrt{\det(\mathbb{G})}}{g_{\theta\theta}} \Gamma^{\phi}_{\theta\theta} \right) \tag{4.10}$$

which, in general, is not constant and, as it can be easily proven, cannot be negative (i.e. $K_G(\theta, \phi) \geqslant 0, \forall(\theta, \phi)$). This leads to the following theorem.

Theorem 4.1 *The manifold surface of an array of N-omnidirectional sensors of arbitrary geometry is never hyperbolic. It is elliptic and embedded in an N-dimensional complex space.*

In Chapter 3 we have stated that surfaces of constant Gaussian curvature are known as "developable surfaces." As $K_G(\theta, \phi)$ of Eq. (4.10) is in general positive or zero, but not constant $\forall(\theta, \phi)$, the array manifold of omnidirectional sensors is not a developable surface. Therefore, the existence of the "development" (see Definition 3.6, page 72) is not guaranteed. However, for some array geometries, the Gaussian curvature of Eq. (4.10) is constant. Array geometries which have been identified to satisfy this condition include all planar arrays as well as a special class of three-dimensional (3D) arrays, known as 3D-grid arrays which are defined and discussed next.

4.2 3D-grid Arrays of Omnidirectional Sensors

Definition 4.1 — *3D-grid arrays.*[1] A three-dimensional array geometry of omnidirectional sensors is said to be a 3D-grid array if and only if the following expression is satisfied:

$$\mathbf{r}\mathbf{r}^T = \rho^2 \mathbb{I}_3 \quad \text{where } \rho \in \mathcal{R} \tag{4.11}$$

Thus, in 3D-grid arrays the vectors $\underline{r}_x, \underline{r}_y$ and $\underline{r}_z$ are not only orthogonal but also have the same magnitude. The following theorem is concerned with the shape of the manifolds of 3D-grid arrays. To prove this theorem it suffices to show that these arrays have constant and positive Gaussian curvature.

Theorem 4.2 *The manifold surface of a 3D-grid array of N-omnidirectional sensors is spherical with radius $\pi\rho$ (and hence developable) embedded in an N-dimensional complex space.*

Example 4.1 The 8-element cube array with all sides equal to one half-wavelength is a 3D-grid array and has a spherical manifold with Gaussian curvature $(2\pi)^{-1}$, or radius $\sqrt{2}\pi$ half-wavelengths.

Table 4.1 summarizes the results of the array manifold parameters for 3D-grid arrays.

Table 4.1 Manifold parameters of 3D-grid arrays.

Intrinsic Parameter	Expression
$\mathbb{G}$ (Eq. (3.5))	$\rho^2\pi^2 \begin{bmatrix} \cos^2\phi, & 0 \\ 0, & 1 \end{bmatrix}$
$\det(\mathbb{G})$	$\rho^4\pi^4\cos^2\phi$
$\mathbf{\Gamma}_{1\theta}$ (Eq. (3.20), ($\varsigma=\theta$))	$\rho^2\pi^2\cos\phi\sin\phi \begin{bmatrix} 0, & -1 \\ 1, & 0 \end{bmatrix}$
$\mathbf{\Gamma}_{1\phi}$ (Eq. (3.20), ($\varsigma=\phi$))	$\rho^2\pi^2\cos\phi\sin\phi \begin{bmatrix} -1, & 0 \\ 0, & 0 \end{bmatrix}$
$\mathbf{\Gamma}_{2\theta}$ (Eq. (3.22), ($\varsigma=\theta$))	$\begin{bmatrix} 0, & -\tan\phi \\ \cos\phi\sin\phi, & 0 \end{bmatrix}$
$\mathbf{\Gamma}_{2\phi}$ (Eq. (3.22), ($\varsigma=\phi$))	$\begin{bmatrix} -\tan\phi, & 0 \\ 0, & 0 \end{bmatrix}$
K_G (Eq. (3.31))	$\dfrac{1}{\rho^2\pi^2}$

[1] 3D-grid arrays are also known as 3D *balance symmetric* arrays.

4.3 Planar Arrays of Omnidirectional Sensors

A planar (2D) array can be seen as a special case of a 3D-array where all the elements of the last (3rd) row of the matrix $\mathbf{r}$ are zeros (i.e. $\underline{r}_z = \underline{0}_N$). In this case, the term $\mathbf{r}\mathbf{r}^T$ has the following form

$$\begin{bmatrix} \mathbb{L}, & \underline{0}_2 \\ \underline{0}_2^T, & 0 \end{bmatrix} \quad \text{where } \mathbb{L} = \begin{bmatrix} c_1, & c_2 \\ c_2, & c_3 \end{bmatrix} \in \mathcal{R}^{2\times 2} \tag{4.12}$$

Furthermore, if $\mathbb{L} = \rho^2 \mathbb{I}_2$ with $\rho \in \mathcal{R}$, the planar array is said to be a *2D-grid array* (or 2D balance symmetric). Fig. 4.1 shows some examples of 2D-grid array geometries.

Theorem 4.3 *The Gaussian curvature K_G of the two-parameter manifold of a planar array of N-omnidirectional sensors is constant at every point and equal to zero.*

Proof. A proof of this theorem is given in Appendix 4.7.1. □

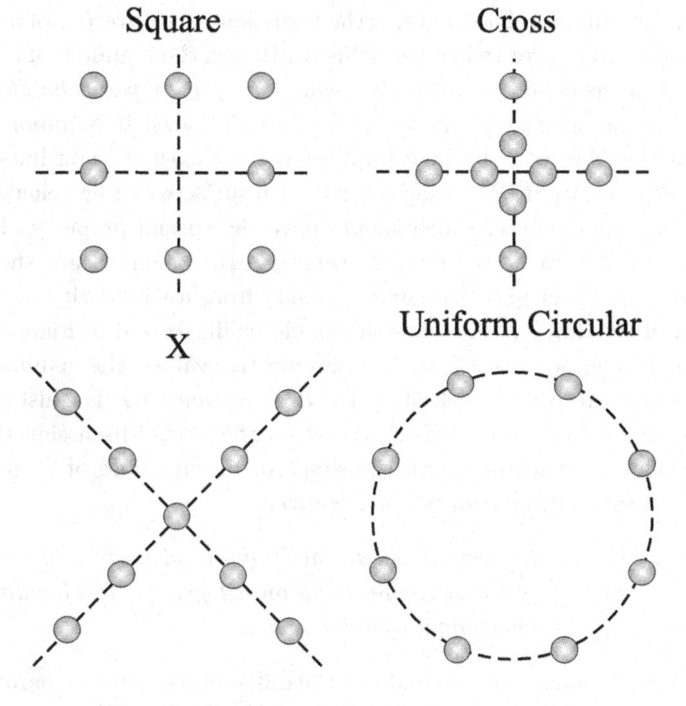

Fig. 4.1 Examples of 2D-grid array geometries.

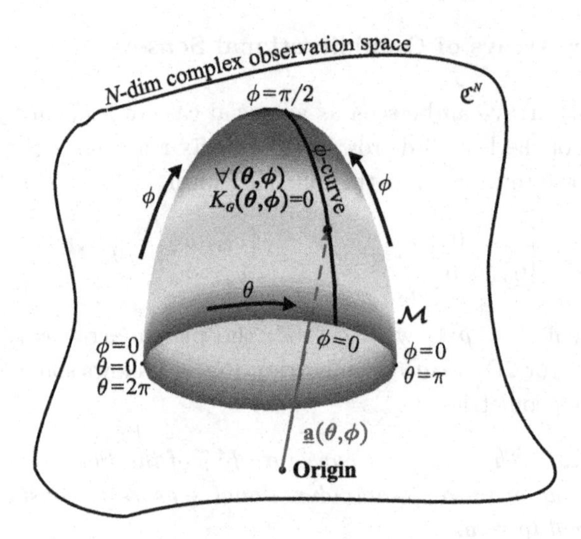

Fig. 4.2 Manifold Surface $\mathcal{M}$ of a Planar Array parametrized in terms of (θ,ϕ). This is a developable manifold-conoid generated by rotating a ϕ-curve (geodesic curve) around the apex.

Thus according to Theorem 4.3, the Gaussian curvature K_G of a planar array is identically zero (wherever defined). Hence, the manifold of a planar array is, what is called in differential geometry, *flat* or *parabolic of conoid shape* with the apex at point $\phi = \pm 90°$ (see Fig. 4.2). It is important to note that the "flatness" does not imply that there exist straight lines, as in the case of a surface in $\mathcal{R}^3$. It means that such surfaces can be generated by rotating a passing curve, which should have the special property of being geodesic, around an apex point. In this case the ϕ-curves are shown to compose a family of geodesic curves passing from a singularity point, i.e. the apex of the developable manifold-conoid as illustrated in Fig. 4.2.

Furthermore, because K_G never takes positive values, the manifold cannot be closed. According to Beltrami's theorem, since K_G is constant, the manifold is a developable surface. Based on the above discussion the following theorem, concerned with the shape of the manifold of an array of omnidirectional sensors, can be easily proven.

Theorem 4.4 *The manifold surface of a planar array of N-omnidirectional sensors is a conoid lying on a hypersphere of radius* $\sqrt{N}$ *embedded in an N-dimensional complex space.*

Table 4.2 summarizes the results of the differential geometry parameters of planar arrays of omnidirectional sensors together with their corresponding simplified expressions for the case of 2D-grid arrays.

Table 4.2 Manifold parameters of planar arrays.

Intrinsic Parameter	2D General Arrays	2D-grid Arrays
$\mathbb{G}$ (Eq. (3.5))	$\begin{bmatrix} g_{\theta\theta}, & g_{\phi\theta} \\ g_{\theta\phi}, & g_{\phi\phi} \end{bmatrix}$	$\rho^2\pi^2 \begin{bmatrix} \cos^2\phi, & 0 \\ 0, & \sin^2\phi \end{bmatrix}$
$\det(\mathbb{G})$	$\frac{\pi^4}{4}\det(\mathbb{L})\sin^2(2\phi)$	$\rho^4\pi^4\cos^2\phi\sin^2\phi$
$\mathbf{\Gamma}_{1\theta}$ (Eq. (3.20), (ζ=θ))	$\begin{bmatrix} \frac{g_{\theta\phi}}{\tan\phi}, & -g_{\theta\theta}\tan\phi \\ \frac{g_{\phi\phi}}{\tan\phi}, & -g_{\theta\phi}\tan\phi \end{bmatrix}$	$\rho^2\pi^2\cos\phi\sin\phi \begin{bmatrix} 0, & -1 \\ 1, & 0 \end{bmatrix}$
$\mathbf{\Gamma}_{1\phi}$ (Eq. (3.20), (ζ=ϕ))	$\begin{bmatrix} -g_{\theta\theta}\tan\phi, & \frac{g_{\theta\phi}}{\tan\phi} \\ -g_{\theta\phi}\tan\phi, & \frac{g_{\phi\phi}}{\tan\phi} \end{bmatrix}$	$\rho^2\pi^2\cos\phi\sin\phi \begin{bmatrix} -1, & 0 \\ 0, & 1 \end{bmatrix}$
$\mathbf{\Gamma}_{2\theta}$ (Eq. (3.22), (ζ=θ))	$\begin{bmatrix} 0, & -\tan\phi \\ \frac{1}{\tan\phi}, & 0 \end{bmatrix}$	$\begin{bmatrix} 0, & -\tan\phi \\ \frac{1}{\tan\phi}, & 0 \end{bmatrix}$
$\mathbf{\Gamma}_{2\phi}$ (Eq. (3.22), (ζ=ϕ))	$\begin{bmatrix} -\tan\phi, & 0 \\ 0, & \frac{1}{\tan\phi} \end{bmatrix}$	$\begin{bmatrix} -\tan\phi, & 0 \\ 0, & \frac{1}{\tan\phi} \end{bmatrix}$
K_G (Eq. (3.31))	0	0

Based on the results presented in Tables 4.1 and 4.2 a number of comments can be made. For instance, the Gaussian curvature of 3D-grid arrays is *always positive* and constant whereas for planar arrays it is *always zero* (constant). This implies that the array manifold of a 3D-grid array is *isometric* with a *sphere* of radius $\rho\pi$ while the manifold of a planar array is *isometric* with a *plane*. In addition, for all 2D-grid and 3D-grid arrays, the manifold metric $\mathbb{G}$ is a diagonal matrix. The determinant of this matrix is given by the 2nd row of Tables 4.1 and 4.2 and can be easily proven for grid arrays. For planar arrays, the proof of $\det(\mathbb{G}) = \frac{\pi^4}{4}\det(\mathbb{L})\sin^2(2\phi)$ is given in Appendix 4.7.2. All these expressions indicate that the determinant of the manifold metric $\mathbb{G}$ is independent of the azimuth angle θ.

4.4 Families of θ- and ϕ-curves on the Manifold Surface

The array manifold can also be described and analyzed by treating this surface as one of the following two families of curves:

(1) the family of θ-parameter curves defined as

$$\{\mathcal{A}_{\theta|\phi_o}, \ \forall\phi_o : \phi_o \in \Omega_\phi\} \tag{4.13}$$

where $\mathcal{A}_{\theta|\phi_o}$ denotes the "θ-curve" on the surface $\mathcal{M}$ corresponding to a constant value of $\phi = \phi_o$ represented as follows:

$$\mathcal{A}_{\theta|\phi_o} = \{\underline{\mathbf{a}}(\theta, \phi_o) \in \mathfrak{C}^N, \ \forall\theta : \theta \in \Omega_\theta, \phi_o = \text{constant}\} \tag{4.14}$$

(2) the family of ϕ-parameter curves defined in a similar fashion as

$$\{\mathcal{A}_{\phi|\theta_o}, \ \forall\theta_o : \theta_o \in \Omega_\theta\} \tag{4.15}$$

where

$$\mathcal{A}_{\phi|\theta_o} = \{\underline{\mathbf{a}}(\theta_o, \phi) \in \mathfrak{C}^N, \ \forall\phi : \phi \in \Omega_\phi, \ \theta_o = \text{constant}\} \tag{4.16}$$

with $\Omega_\theta \triangleq [0^\circ \text{ to } 360^\circ)$ and $\Omega_\phi \triangleq [0^\circ \text{ to } 90^\circ)$. Note that

$$\mathcal{M} = \{\mathcal{A}_{\theta|\phi_o}, \ \forall\phi_o : \phi_o \in \Omega_\phi\} = \{\mathcal{A}_{\phi|\theta_o}, \ \forall\theta_o : \theta_o \in \Omega_\theta\} \tag{4.17}$$

Figure 4.3 shows an illustrative representation of one "θ-curve" and one "ϕ-curve" lying on the array manifold surface $\mathcal{M}$.

These two curves lying on $\mathcal{M}$ can be analyzed using the geodesic curvature. The geodesic curvature is one of the most important parameters of the intrinsic geometry of a surface. In particular this parameter is associated with curves lying on the surface and remains invariant under an "isometry."

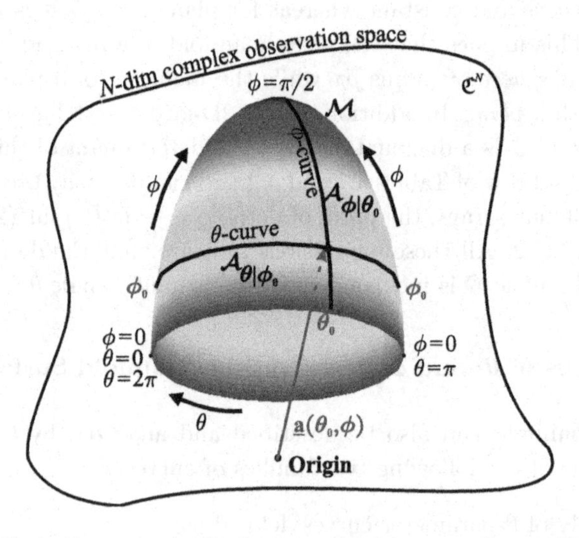

Fig. 4.3 An illustrative example of a θ-curve and a ϕ-curve on the manifold surface $\mathcal{M}$.

It can be proven that all 3D and all planar arrays have zero geodesic curvature for the ϕ-curves. This implies that

$$\boxed{\kappa_{g,\phi} = 0 \Longleftrightarrow \phi\text{-curves} = \text{geodesic curves}} \tag{4.18}$$

Furthermore, the geodesic curvature $\kappa_{g,\theta}$ for the θ-curves for 3D-grid and 2D-grid arrays is constant (depending on ϕ_o) but is not zero. Hence

$$\boxed{\kappa_{g,\theta} \neq 0 \Longleftrightarrow \theta\text{-curves} \neq \text{geodesic curves}} \tag{4.19}$$

while constant $\kappa_{g,\theta}$ implies that the development of the θ-curves of grid arrays are circles whose radius is a function of ϕ_o.

Using Eq. (3.42) it can be proven that for a general planar array the geodesic curvature $\kappa_{g,\theta}$ is given as follows

$$\kappa_{g,\theta} = \sqrt{\frac{\det(\mathbb{G})}{g_{\theta\theta}^3}} \frac{1}{\tan\phi_o} \tag{4.20}$$

and this may not even be constant. The above results are summarized in Table 4.3.

Table 4.3 Geodesic curvature.

Curve	κ_g	3D-grid Arrays	Planar Arrays	2D-grid Arrays
$\mathcal{A}_{\theta\vert\phi_o}$	$\kappa_{g,\theta}$	$\frac{\tan\phi_o}{\rho\pi}$	$\sqrt{\frac{\det(\mathbb{G})}{g_{\theta\theta}^3}} \frac{1}{\tan\phi_o}$	$\frac{1}{\rho\pi\cos\phi_o}$
$\mathcal{A}_{\phi\vert\theta_o}$ (hyperhelix)	$\kappa_{g,\phi}$	0	0	0

For the 2D-grid and 3D-grid arrays, the off-diagonal elements of the manifold metric $\mathbb{G}$ (i.e. $g_{\theta\phi}$, $g_{\phi\theta}$) are zero (see Tables 4.1, 4.2) and hence the θ-curves and ϕ-curves are orthogonal and constitute geodesic coordinates i.e.

$$\left.\begin{array}{l}(\theta\text{-curves}) \perp (\phi\text{-curves}) \\ (\phi\text{-curves}) = \text{geodesic curve}\end{array}\right\} \Longrightarrow \textit{geodesic}\text{ coordinates} \tag{4.21}$$

Figure 4.4 shows an illustrative representation of a planar array manifold surface and its families of θ- and ϕ-curves as geodesic coordinates.

Theorem 4.5 *The geodesic curves (ϕ-curves) on the manifold surface are complex hyperhelices embedded in $\mathfrak{C}^N$.*

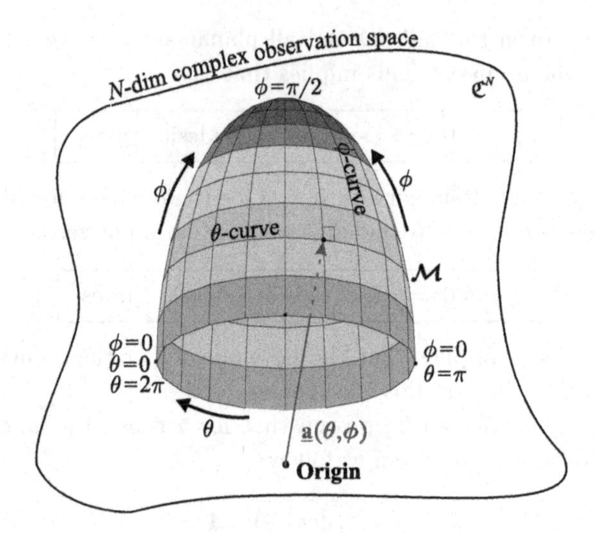

Fig. 4.4 Families of θ-curves and ϕ-curves on the manifold $\mathcal{M}$ of a planar array.

Proof. Eq. (4.1), for constant θ (of value θ_0), can be expressed as

$$\begin{aligned}\underline{\mathbf{a}}(\theta_o, \phi) &= \exp(-j\pi(\underline{r}_x \cos\theta_o + \underline{r}_y \sin\theta_o)\cos\phi) \\ &= \exp(-j\pi\underline{r}(\theta_o)\cos\phi)\end{aligned} \tag{4.22}$$

where

$$\underline{r}(\theta_o) = \underline{r}_x \cos\theta_o + \underline{r}_y \sin\theta_o \tag{4.23}$$

which matches Eq. (2.27) for $\underline{r} = \underline{r}(\theta_o)$, $p = \phi$ and $\underline{v} = \underline{0}$. However, hyperhelical curves, such as ϕ-curves, have all their curvatures constant, independent of the parameters (θ, ϕ), as shown in Chapter 2. $\square$

Hence hyperhelical curves on the manifold surface $\mathcal{M}$ are not only analytically "convenient" but also geodesic curves. Overall it may be concluded that, although both families of curves can be used to describe the manifold surface $\mathcal{M}$, the θ-curves are the more complicated of the two families.

Corollary 4.1 *All members of the family of ϕ-curves of planar arrays are identical apart from their length l_m and position in complex N-dimensional space.*

It is immediately apparent that

$$\underline{\mathbf{a}}(\phi \mid \theta_o) = \underline{\mathbf{a}}(\phi \mid \theta_o + 180°) \tag{4.24}$$

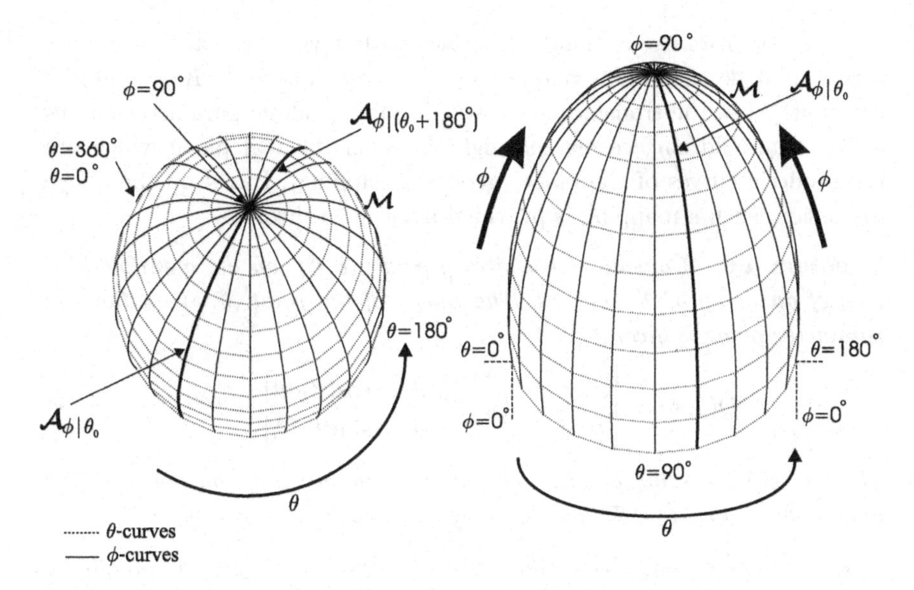

Fig. 4.5 Illustration of *composite curves* on the array manifold surface.

which implies that the two ϕ-curves at $\theta = \theta_o$ and $\theta = \theta_o + 180°$ have the same length and curvatures. As illustrated in Fig. 4.5, these two ϕ-curves can be considered as a continuation of one another, forming a *composite* ϕ-curve having a hyperhelical shape of double the length of the original curves. That is

$$\mathcal{A}_{\phi|\theta_o} + \mathcal{A}_{\phi|\theta_o+180°} = \text{composite curve} \tag{4.25}$$

$$\begin{aligned}(l_m \text{ of } \mathcal{A}_{\phi|\theta_o}) &= (l_m \text{ of } \mathcal{A}_{\phi|\theta_o+180°}) \\ &= (s(90°), \text{ Eq. (2.28)}) = \pi \left\|\underline{r}(\theta_o)\right\|\end{aligned} \tag{4.26}$$

$$(l_m \text{ of } \textit{composite} \text{ curve}) = 2\pi \left\|\underline{r}(\theta_o)\right\| \tag{4.27}$$

4.5 "Development" of Non-linear Array Geometries

In Section 3.5, "development" has been described as a local mapping which represents a complicated surface (embedded in a multidimensional complex space) as a simpler surface. This representation would simplify the analysis of array manifold surfaces and consequently of array systems. The simplest surface, from a conceptual point of view, is the real plane. In order to be useful, this mapping should preserve the main characteristics of the manifold surface. For instance, a manifold surface with a varying Gaussian curvature cannot be adequately mapped onto a plane. Therefore K_G should be constant. Also, the transformation should map geodesic parameter curves on the manifold surface onto straight lines on the real plane which are the geodesic curves of the plane (geodesic mapping). Based on the above discussion, such a mapping is provided below as a theorem.

Theorem 4.6 *Consider an arbitrary point, (θ, ϕ), on the manifold surface of an array of N sensors. The image of the point (θ, ϕ) on the real parameter plane is given by*

$$\underline{D}(\theta, \phi) = \begin{bmatrix} x_0(\phi) + \int_0^\theta \kappa_{g,\theta}^{-1}(\vartheta, \phi)\cos(\vartheta)d\vartheta \\ y_0(\phi) + \int_0^\theta \kappa_{g,\theta}^{-1}(\vartheta, \phi)\sin(\vartheta)d\vartheta \end{bmatrix} \in \mathcal{R}^2 \tag{4.28}$$

where $\kappa_{g,\theta}(\theta, \phi)$ is the geodesic curvature of the θ-curves on the manifold and $x_0(\phi)$ and $y_0(\phi) \in \mathcal{R}$ are the integration constants such that:

$$x_0 = -\tfrac{1}{2}\int_0^\pi \kappa_{g,\theta}^{-1}(\vartheta, \phi)\cos(\vartheta)d\vartheta \quad \textit{and} \quad y_0 = -\tfrac{1}{2}\int_0^\pi \kappa_{g,\theta}^{-1}(\vartheta, \phi)\sin(\vartheta)d\vartheta$$

The locus of the images $\underline{D}(\theta, \phi)\, \forall(\theta, \phi)$ is said to be the development[2] *of the manifold and has the following properties:*

(i) It exists if and only if the Gaussian curvature of the manifold surface is constant.

(ii) The arc length $s_d(\theta, \phi_o)$ of θ-curves on the development is given by

$$s_d(\theta, \phi_o) = \int_p \kappa_{g,\theta}^{-1}(\vartheta, \phi_o)\, d\vartheta \tag{4.29}$$

(iii) The curvature of the θ-curves on the development is equal to the geodesic curvature $\kappa_{g,\theta}(\theta, \phi_o)$ of the θ-curves on the manifold surface.

Proof. The proof is given in Appendix 4.7.3. □

Note that the expressions for the integration constants assume the property $\kappa_{g,\theta}(\theta, \phi_o) = \kappa_{g,\theta}(\theta + \pi, \phi_o)$ (which is generally valid), and have the effect of shifting the curves to a common centre, i.e.

$$\underline{D}(\theta, \phi) + \underline{D}(\theta + \pi, \phi) = \underline{0} \tag{4.30}$$

The first condition of the theorem (i.e. $K_G = \textit{constant}$) is valid for 3D-grid and all planar arrays, which implies that these arrays are "developable," with their "development" expressions summarized in Table 4.4.

---

[2] More on "development" and "developables" can be found in [17].

Table 4.4 Development of Grid and Planar Arrays

3D-grid	$\underline{D}(\theta,\phi) = \frac{\rho\pi}{\tan\phi}\begin{bmatrix}\sin\theta \\ -\cos\theta\end{bmatrix}$
Planar	$\underline{D}(\theta,\phi) = \frac{\pi\cos\phi}{\sqrt{\det(\mathbb{L})}}\begin{bmatrix}\int_0^\theta f(\vartheta)\cos\vartheta d\vartheta - \frac{1}{2}\int_0^\pi f(\vartheta)\cos\vartheta d\vartheta \\ \int_0^\theta f(\vartheta)\sin\vartheta d\vartheta - \frac{1}{2}\int_0^\pi f(\vartheta)\sin\vartheta d\vartheta\end{bmatrix}$ with $\mathbb{L} = \begin{bmatrix}\overbrace{\|\underline{r}_x\|^2}^{\triangleq c_1}, & \overbrace{\underline{r}_x^T\underline{r}_y}^{\triangleq c_2} \\ \underbrace{\underline{r}_x^T\underline{r}_y}_{\triangleq c_2}, & \underbrace{\|\underline{r}_y\|^2}_{\triangleq c_3}\end{bmatrix}$ and $f(\vartheta) \triangleq \sqrt{(c_3\cos^2\vartheta - 2c_2\cos\vartheta\sin\vartheta + c_1\sin^2\vartheta)^3}$
2D-grid	$\underline{D}(\theta,\phi) = \rho\pi\cos\phi\begin{bmatrix}\sin\theta \\ -\cos\theta\end{bmatrix}$

Example 4.2 Figures 4.6, 4.7 and 4.8 provide the development of representative 3D-grid, planar and 2D-grid array geometries. For instance the development of a 3D-grid array with sensors located on the eight vertices of a cube of side one half-wavelength and with reference point the centre of the cube (array centroid) is shown in Fig. 4.6. For the planar array with the positions of the sensors, in half-wavelengths, as given below:

$$\mathbf{r} = \begin{bmatrix}-2.1 & -1.1 & 0.4 & 0.9 & 1.9 \\ -2.4 & 1.1 & 0.6 & -0.4 & 1.1 \\ 0 & 0 & 0 & 0 & 0\end{bmatrix} \tag{4.31}$$

the development is shown Fig. 4.7, while the development for a 5-sensor uniform circular array (2D-grid array) of radius one half wavelength is presented in Fig. 4.8.

It is clear that the manifold $\mathcal{M}$ of a planar array is mapped on the parameter plane (θ, ϕ) as a system of homocentric ellipses while if the array is a grid array (2D-grid or 3D-grid) this mapping has a circular shape and is "geodesic" mapping. In the case of 2D this is also an "isometric" mapping. In other words the grid arrays have circular development for the θ-curves implying that the geodesic curvature of these curves is constant. The development of the ϕ-curves, which are geodesic curves on the surface, are straight lines. The development of θ-curves for general planar arrays

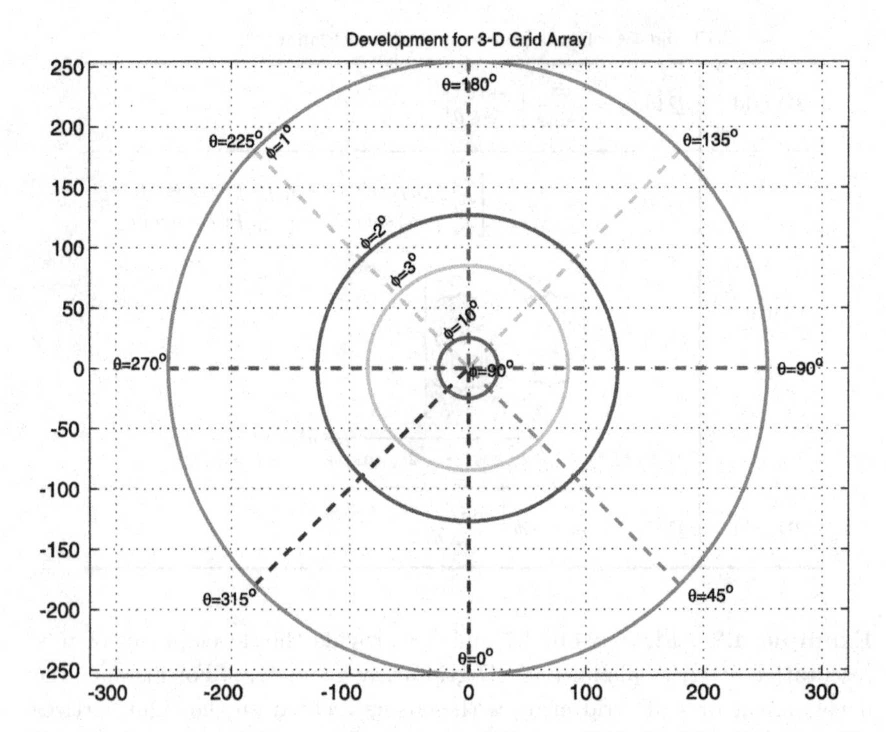

Fig. 4.6 θ-curves and ϕ-curves development of the manifold of a 3D-grid array with sensors located on the eight vertices of a cube of side one half-wavelength and with reference point the centre of the cube (array centroid).

have an ellipsoidal shape with the two main axes at angles ψ_0 and $\psi_0 + \pi/2$ with respect to the x-axis where:

$$\psi_0 = \frac{1}{2}\tan^{-1}\left(\frac{2c_2}{c_1 - c_3}\right) \tag{4.32}$$

with c_i denoting the elements of the matrix $\mathbb{L}$, i.e. $\mathbb{L} = \begin{bmatrix} c_1, & c_2 \\ c_2, & c_3 \end{bmatrix}$. Indeed the angle ψ_0 can be found by examining the points of extremum curvatures. Since the curvature of the development is equal to $\kappa_{g\theta}$, the geodesic curvature of the θ-curves of the planar array manifold is differentiated with respect to θ and equated to zero:

$$\frac{d\kappa_{g\theta}}{d\theta} = 0 \tag{4.33}$$

$$\Leftrightarrow \frac{d(c_3\cos^2\psi_0 - 2c_2\cos\psi_0\sin\psi_0 + c_1\sin^2\psi_0)^{-3/2}}{d\psi_0} = 0$$

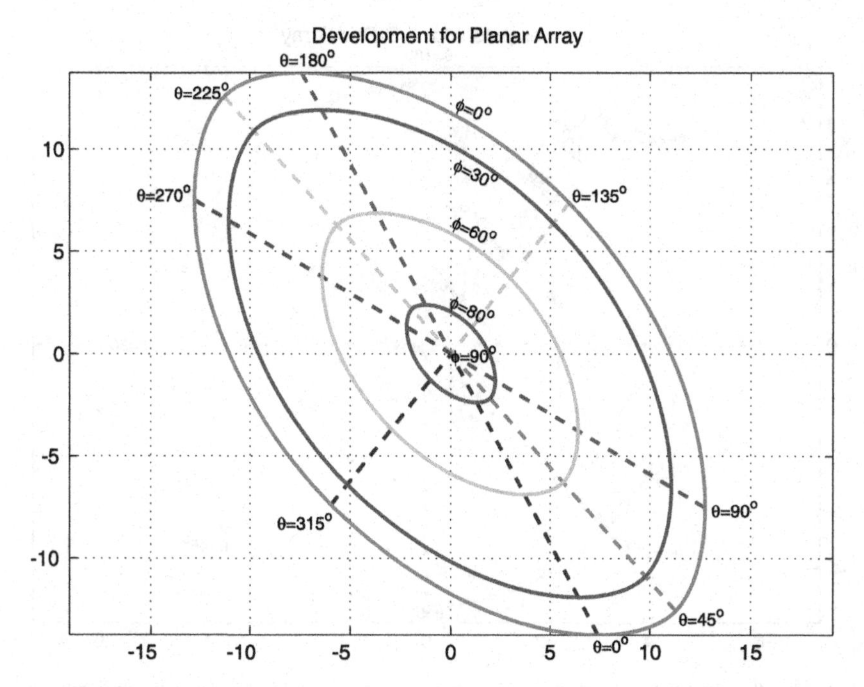

Fig. 4.7 θ-curves and ϕ-curves development of the manifold of a planar array of 5-sensor with positions given by Eq. (4.31).

$$\Leftrightarrow (c_3 \cos^2 \psi_0 - 2c_2 \cos \psi_0 \sin \psi_0 + c_1 \sin^2 \psi_0)^{-5/2}$$
$$\times((c_1 - c_3)\sin(2\psi_0) - 2c_2 \cos(2\psi_0)) = 0$$

to yield two equations as follows:

$$\Leftrightarrow \begin{cases} c_3 \cos^2 \psi_0 - 2c_2 \cos \psi_0 \sin \psi_0 + c_1 \sin^2 \psi_0 = 0 \\ \text{or} \\ (c_1 - c_3)\sin(2\psi_0) - 2c_2 \cos(2\psi_0) \qquad = 0 \end{cases} \tag{4.34}$$

Considering the first equation of Eq. (4.34):

$$c_3 \cos^2 \psi_0 - 2c_2 \cos \psi_0 \sin \psi_0 + c_1 \sin^2 \psi_0 = 0$$

the discriminant gives $-4\det(\mathbb{L}) < 0$. Hence the first equation has no solution. The second equation gives

$$\tan(2\psi_0) = \frac{2c_2}{c_1 - c_3} \quad \text{and hence} \quad \psi_0 = \frac{1}{2}\tan^{-1}\left(\frac{2c_2}{c_1 - c_3}\right) \tag{4.35}$$

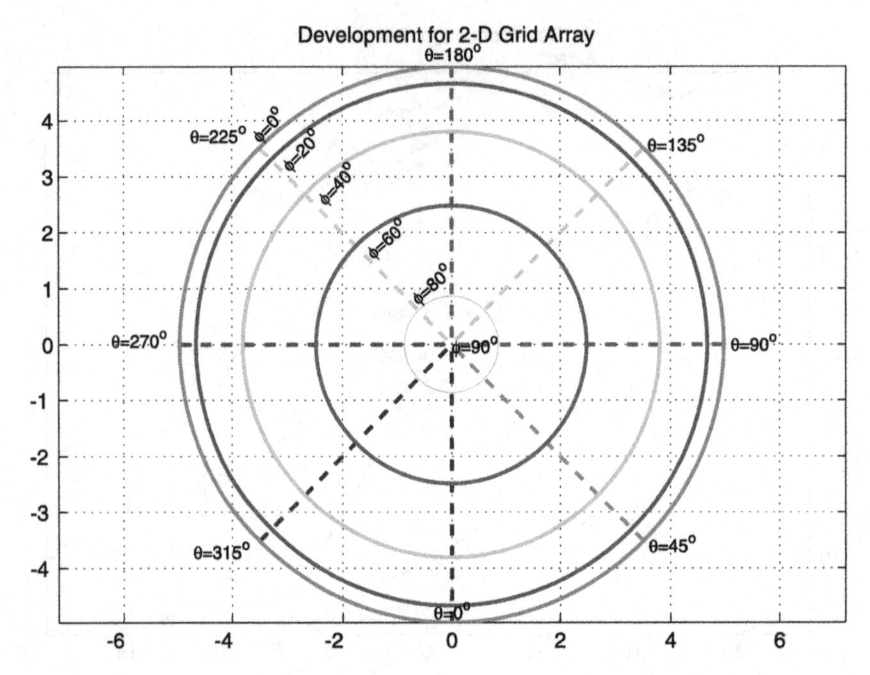

Fig. 4.8 θ-curves and ϕ-curves development of the manifold of a 5-sensor uniform circular array (2D-grid array) of radius one half-wavelength.

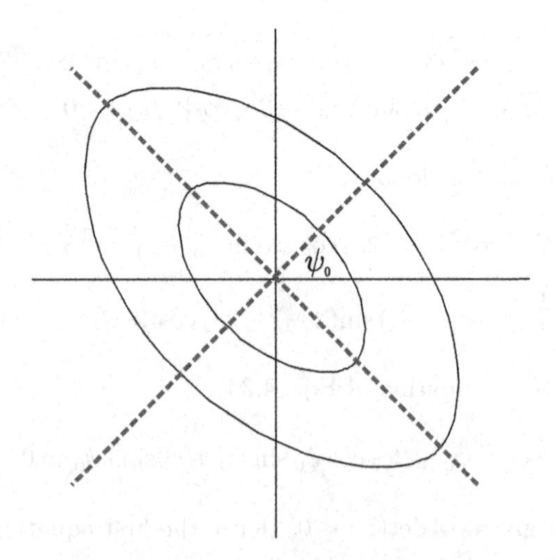

Fig. 4.9 Orientation of the "development" of θ-curves — axes at angles ψ_0 and $\psi_0+90°$.

4.6 Summary

A number of differential geometry parameters have been estimated for 3D and planar arrays of omnidirectional sensors parametrized in terms of azimuth and elevation angles. The main results are summarized as follows:

- The manifolds of 3-dimensional arrays of arbitrary geometry are locally elliptic ($K_G \geqslant 0, \forall(\theta,\phi)$),
- The manifolds of 3D-grid arrays are spherical ($K_G = \text{constant}, \forall(\theta,\phi)$),
- All planar arrays have a manifold of conoidal shape ($K_G = 0, \forall(\theta,\phi)$),
- The family of the ϕ-curves is a family of geodesic curves for all grid and planar arrays,
- The θ-curves are not geodesic curves,
- The ϕ-curves have a hyperhelical shape,
- θ-curves and the ϕ-curves constitute a system of *geodesic coordinates*,
- For planar and grid arrays the development of ϕ-curves are straight lines while the development of θ-curves have an ellipsoidal shape.

4.7 Appendices

4.7.1 *Proof that the Gaussian Curvature of an Omni-directional Sensor Planar Array Manifold is Zero*

Equation (4.10) gives the expression for the Gaussian curvature of an array of omnidirectional sensors, which can be written in an equivalent way as follows:

$$\begin{aligned} K_G &= \frac{1}{\sqrt{\det(\mathbb{G})}}\frac{1}{g_{\theta\theta}^2\tan^2\phi} \\ &\quad\times\left(\begin{array}{c} \frac{1}{2}\left(g_{\theta\theta}\frac{dg_{\phi\phi}}{d\phi}+g_{\phi\phi}\frac{dg_{\theta\theta}}{d\phi}-2g_{\theta\phi}\frac{dg_{\theta\phi}}{d\phi}\right) \\ \times(\sqrt{\det(\mathbb{G})})^{-1}g_{\theta\theta}\tan\phi \\ -\sqrt{\det(\mathbb{G})}\left(\frac{dg_{\theta\theta}}{d\phi}\tan\phi+g_{\theta\theta}(1+\tan^2\phi)\right) \end{array}\right) \\ &= \frac{g_{\theta\theta}\frac{dg_{\phi\phi}}{d\phi}+g_{\phi\phi}\frac{dg_{\theta\theta}}{d\phi}-2g_{\theta\phi}\frac{dg_{\theta\phi}}{d\phi}}{2\det(\mathbb{G})g_{\theta\theta}\tan\phi} \\ &\quad-\frac{\frac{dg_{\theta\theta}}{d\phi}\tan\phi+g_{\theta\theta}(1+\tan^2\phi)}{g_{\theta\theta}^2\tan^2\phi} \end{aligned} \tag{4.36}$$

By using the Christoffel symbols of the second kind, $\Gamma^{\theta}_{\theta\phi}$ and $\Gamma^{\phi}_{\phi\phi}$, which can be expressed in terms of the elements of the manifold metric $\mathbb{G}$ as

$$\left\{\begin{array}{l} \Gamma^{\theta}_{\theta\phi} = \frac{1}{2\det(\mathbb{G})}\left(g_{\phi\phi}\frac{dg_{\theta\theta}}{d\phi} - g_{\theta\phi}\frac{dg_{\phi\phi}}{d\phi}\right) \\ \Gamma^{\phi}_{\phi\phi} = \frac{1}{2\det(\mathbb{G})}\left(g_{\theta\theta}\frac{dg_{\phi\phi}}{d\phi} + g_{\theta\phi}\frac{dg_{\phi\phi}}{d\phi} - 2g_{\theta\phi}\frac{dg_{\theta\phi}}{d\phi}\right) \end{array}\right\} \tag{4.37}$$

Eq. (4.36) can be rewritten as

$$K_G = \frac{\Gamma^{\theta}_{\theta\phi} + \Gamma^{\phi}_{\phi\phi}}{g_{\theta\theta}\tan\phi} - \frac{\frac{dg_{\theta\theta}}{d\phi}\tan\phi + g_{\theta\theta}\left(1+\tan^2\phi\right)}{g^2_{\theta\theta}\tan^2\phi} \tag{4.38}$$

Using the previous expression in conjunction with Table 4.2, which provides the Christoffel symbols $\Gamma^{\theta}_{\theta\phi}$ and $\Gamma^{\phi}_{\phi\phi}$ for a planar array as a function of ϕ, we have

$$\begin{aligned} K_G &= \frac{-\tan\phi + \frac{1}{\tan\phi}}{g_{\theta\theta}\tan\phi} - \frac{\frac{dg_{\theta\theta}}{d\phi}\tan\phi + g_{\theta\theta}\left(1+\tan^2\phi\right)}{g^2_{\theta\theta}\tan^2\phi} \\ &= -\frac{\frac{dg_{\theta\theta}}{d\phi} + 2g_{\theta\theta}\tan\phi}{g^2_{\theta\theta}\tan\phi} \end{aligned} \tag{4.39}$$

By using the property:

$$\ddot{\underline{k}}_{\theta\phi} = -\dot{\underline{k}}_{\theta}\tan\phi \tag{4.40}$$

K_G is simplified to

$$K_G = -\frac{-2g_{\theta\theta}\tan\phi + 2g_{\theta\theta}\tan\phi}{g^2_{\theta\theta}\tan\phi} = 0 \tag{4.41}$$

□

4.7.2 *Proof of the Expression of* det($\mathbb{G}$) *for Planar Arrays in Table 4.2*

For planar arrays, the term $\mathbf{r}\mathbf{r}^T$ has the following form $\begin{bmatrix} \mathbb{L}, & \underline{0}_2 \\ \underline{0}_2^T, & 0 \end{bmatrix}$ where $\mathbb{L} = \begin{bmatrix} c_1, & c_2 \\ c_2, & c_3 \end{bmatrix} \in \mathcal{R}^{2\times 2}$ and the elements of the manifold metric $\mathbb{G}$ for an array of omnidirectional sensors can be explicitly written as a function of the elements of $\mathbb{L}$ as

$$\left\{\begin{array}{l} g_{\theta\theta} = \pi^2\cos^2\phi\left(c_3\cos^2\theta - 2c_2\cos\theta\sin\theta + c_1\sin^2\theta\right) \\ g_{\phi\phi} = \pi^2\sin^2\phi\left(c_1\cos^2\theta + 2c_2\cos\theta\sin\theta + c_3\sin^2\theta\right) \\ g_{\theta\phi} = \pi^2\cos\phi\sin\phi\left(c_2\left(\sin^2\theta - \cos^2\theta\right) + (c_1 - c_3)\cos\theta\sin\theta\right) \end{array}\right. \tag{4.42}$$

Using Eq. (4.42) the determinant of the manifold metric $\mathbb{G}$ is found to be

$$\det(\mathbb{G}) = \pi^4 \cos^2\phi \sin^2\phi \begin{pmatrix} (c_3\cos^2\theta - 2c_2\cos\theta\sin\theta + c_1\sin^2\theta) \\ \times(c_1\cos^2\theta + 2c_2\cos\theta\sin\theta + c_3\sin^2\theta) \\ -(c_2(\sin^2\theta - \cos^2\theta) + (c_1 - c_3)\cos\theta\sin\theta)^2 \end{pmatrix}$$

Let the first term inside the braces be denoted by X and the second by Y, i.e.

$$\begin{aligned} X &= \begin{matrix} (c_3\cos^2\theta - 2c_2\cos\theta\sin\theta + c_1\sin^2\theta) \\ \times(c_1\cos^2\theta + 2c_2\cos\theta\sin\theta + c_3\sin^2\theta) \end{matrix} \\ Y &= (c_2(\sin^2\theta - \cos^2\theta) + (c_1 - c_3)\cos\theta\sin\theta)^2 \end{aligned} \tag{4.43}$$

The above expressions can be expanded to

$$\begin{cases} X = c_1c_3\chi^4 - 2c_1c_2\chi^3\sigma + c_1^2\chi^2\sigma^2 \\ \quad + 2c_2c_3\chi^3\sigma - 4c_2^2\chi^2\sigma^2 + 2c_1c_2\chi\sigma^3 \\ \quad + c_3^2\chi^2\sigma^2 - 2c_2c_3\chi\sigma^3 + c_1c_3\sigma^4 \\ Y = c_2^2\chi^4 - c_1c_2\chi^3\sigma + c_2c_3\chi^3\sigma \\ \quad - c_2^2\chi^2\sigma^2 - c_1c_2\chi^3\sigma + c_1^2\chi^2\sigma^2 \\ \quad - c_1c_3\chi^2\sigma^2 + c_1c_2\chi\sigma^3 + c_2c_3\chi^3\sigma \\ \quad - c_1c_3\chi^2\sigma^2 + c_3^2\chi^2\sigma^2 - c_2c_3\chi\sigma^3 \\ \quad - c_2^2\chi^2\sigma^2 + c_1c_2\chi\sigma^3 - c_2c_3\chi\sigma^3 + c_2^2\sigma^4 \end{cases}$$

where $\chi = \cos\theta$ and $\sigma = \sin\theta$ and hence $\chi^2 + \sigma^2 = 1$.

Hence:

$$\begin{aligned} \det(\mathbb{G}) &= \pi^4\cos^2\phi\sin^2\phi(X - Y) \\ &= \pi^4\cos^2\phi\sin^2\phi \\ &\quad \times((c_1c_3 - c_2^2)\chi^4 + 2(c_1c_3 - c_2^2)\chi^2\sigma^2 + (c_1c_3 - c_2^2)\sigma^4) \\ &= \pi^4\cos^2\phi\sin^2\phi(c_1c_3 - c_2^2)(\chi^2 + \sigma^2)^2 \\ &= \pi^4\cos^2\phi\sin^2\phi(c_1c_3 - c_2^2) \\ &= \frac{\pi^4}{4}\det(\mathbb{L})\sin^2(2\phi) \end{aligned}$$

4.7.3 *Proof of "Development" Theorem 4.6*

(i) According to Beltrami's theorem [20, 21], because the plane has a constant Gaussian curvature (equal to zero), the condition for the existence of a distortionless mapping is that the array manifold surface should have a constant Gaussian curvature.

(ii) Considering the development of the manifold θ-curves for a constant ϕ_o, $\underline{D}(\theta, \phi_o)\ \forall\theta$, its arc length is given by

$$s_d(\theta, \phi_o) = \int_\theta \left\| \frac{d\underline{D}(\vartheta, \phi_o)}{d\vartheta} \right\| d\vartheta \tag{4.44}$$

Using the Leibnitz rule of integral differentiation on Eq. (4.28):

$$\frac{d\underline{D}(\theta, \phi_o)}{d\theta} = \kappa_g^{-1}(\theta, \phi_0) \begin{bmatrix} \cos\theta \\ \sin\theta \end{bmatrix}$$

and hence $\left\| \frac{d\underline{D}(\theta,\phi_o)}{d\theta} \right\| = \kappa_{g\theta}^{-1}(\theta, \phi_o)$ proving Eq. (4.29).

(iii) The curvature of a curve is the magnitude of the derivative of the normalized tangent vector with respect to its arc length. Hence, considering the development of the manifold θ-curves for a constant ϕ_o, let $\underline{u}(\theta)$ be its normalized tangent vector given by

$$\underline{u}(\theta, \phi_o) = \frac{\frac{d\underline{D}(\theta,\phi_o)}{d\theta}}{\left\| \frac{d\underline{D}(\theta,\phi_o)}{d\theta} \right\|} = \pm \begin{bmatrix} \cos\theta \\ \sin\theta \end{bmatrix} \tag{4.45}$$

Differentiating Eq. (4.29) leads to

$$\frac{ds_d}{d\theta} = \kappa_g^{-1}(\theta, \phi_o) \tag{4.46}$$

and hence,

$$\frac{d\theta}{ds_d} = \kappa_g(\theta, \phi_0) \tag{4.47}$$

The magnitude of the derivative of the normalized tangent vector with respect to its arc length hence becomes

$$\begin{aligned} \left\| \frac{d\underline{u}\,(\theta, \phi_o)}{ds_d} \right\| &= \left\| \frac{d\underline{u}\,(\theta, \phi_o)}{d\theta} \frac{d\theta}{ds_d} \right\| \\ &= \left\| \begin{bmatrix} -\sin\theta \\ \cos\theta \end{bmatrix} \kappa_g\,(\theta, \phi_o) \right\| = \kappa_g(\theta, \phi_o) \end{aligned} \tag{4.48}$$

The curvature of the development is thus equal to the geodesic curvature of the manifold curve.

Chapter 5

Non-Linear Arrays: (α, β)-Parametrization

In the previous chapter the array manifold surface $\mathcal{M}$ was defined in terms of azimuth θ and elevation ϕ angles and its intrinsic geometrical properties were highlighted. Furthermore, the surface $\mathcal{M}$ was also considered based on the families of θ-curves and ϕ-curves, whose characteristics were examined using the differential geometry framework presented in Chapters 2 and 3. In Chapter 4 it was observed that while the ϕ-curves were hyperhelical and geodesic, the θ-curves possessed neither property. It was also pointed out that since the θ-curves are not hyperhelical, their curvatures depend on θ and so analytical evaluation of curvatures of order greater than two can become exceedingly laborious and impractical.

In spite of the intuitive appeal of using azimuth and elevation angles as the medium of array parametrization, this choice of angles is by no means unique. In this chapter an alternative parametrization, known as "cone-angle", is presented. This parametrization is particularly suitable for planar arrays resulting in two families of hyperhelical curves which can provide a great deal of additional insight into the nature of planar array behavior.

5.1 Mapping from the (θ, ϕ) Parameter Space to Cone-Angle Parameter Space

For an array of N-omnidirectional sensors with positions given by the $3 \times N$ matrix $\mathbf{r} = [\underline{r}_x, \underline{r}_y, \underline{r}_z]^T$, using the directional parameters (θ, ϕ) the array manifold $\mathcal{M}$ was defined in the previous chapter as

$$\mathcal{M} = \{\underline{\mathbf{a}}(\theta, \phi) \in \mathfrak{C}^N, \ \forall(\theta, \phi) : (\theta, \phi) \in \Omega\} \tag{5.1}$$

where

$$\Omega = \{(\theta, \phi) : \theta \in [0°, 360°) \quad \text{and} \quad \phi \in [0°, 90°)\}$$

with

$$\underline{\mathbf{a}}(\theta, \phi) = \exp(-j\mathbf{r}^T\underline{\mathbf{k}}(\theta, \phi)) \tag{5.2}$$

An alternative parametrization of $\mathcal{M}$ is based on two new angles, known as the cone angles α-β, and defined as follows. If the x- and y-axes of the Cartesian frame (x-y-z) are rotated on the (x, y) plane around the z-axis by an angle Θ, then the frame ($\check{x}$-$\check{y}$-z) is obtained. The cone-angle α is defined as the angle between the wavenumber vector $\underline{\mathbf{k}}(\theta, \phi)$ and the positive side of the $\check{x}$-axis, while β is the angle between $\underline{\mathbf{k}}(\theta, \phi)$ and the positive side of $\check{y}$-axis.

The relation between the two conventions/parametrization is

$$\cos\alpha = \cos\phi\cos(\theta - \Theta) \quad 0° \leq \alpha \leq 180° \tag{5.3}$$

$$\cos\beta = \cos\phi\sin(\theta - \Theta) \quad 0° \leq \beta \leq 180° \tag{5.4}$$

and is illustrated in Fig. 5.1. These angles α and β are called cone-angles because the loci of wavenumber vectors of constant $\alpha = \alpha_0$ (or equivalently constant $\beta = \beta_0$) form cones about the $\check{x}$ (equivalently the $\check{y}$) axis. In other words, all possible incident signals from a fixed angle α, or β, form a cone in space. Fig. 5.2 shows this for a random choice of cone-angles α_0 and β_0.

The mapping

$$(\theta, \phi) \longmapsto (\alpha, \beta) \tag{5.5}$$

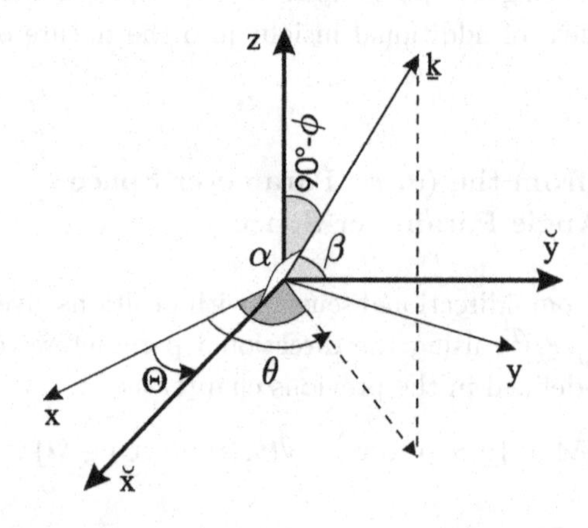

Fig. 5.1 Azimuth-elevation and cone-angles conventions.

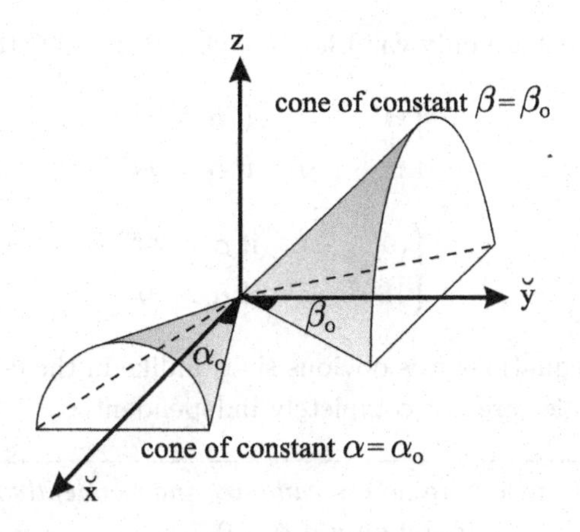

Fig. 5.2 Half-cones formed by wavenumber vectors of constant α and constant β.

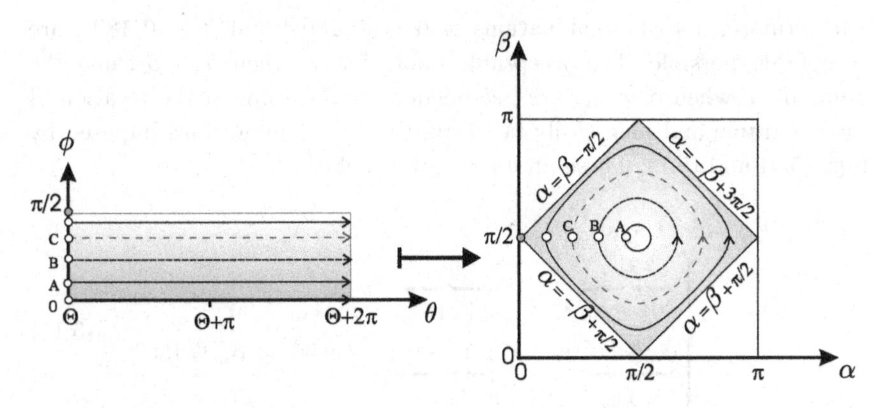

Fig. 5.3 Mapping from the (θ, ϕ) parameter space on to the (α, β) parameter space.

is one-to-one and an illustrative representation of this mapping is given in Fig. 5.3. For a given pair of values of cone-angles α and β and a given rotation Θ of the x-y axes, the expressions that calculate θ and ϕ can be derived from Eqs. (5.3) and (5.4) and are given by

$$\theta = \Theta + \arctan\left(\frac{\cos\beta}{\cos\alpha}\right) \tag{5.6}$$

$$\begin{aligned}\phi &= \arccos\left(\frac{\cos\beta}{\sin(\theta - \Theta)}\right) \\ &= \arccos\left(\frac{\cos\beta}{\sin\left(\arctan\left(\frac{\cos\beta}{\cos\alpha}\right)\right)}\right)\end{aligned} \tag{5.7}$$

These equations are only valid for $\beta \neq 90°$. If $\beta = 90°$ then

$$\theta = \begin{cases} \Theta & \text{if } \alpha \leq 90° \\ \Theta + 180° & \text{if } \alpha > 90° \end{cases} \tag{5.8}$$

$$\phi = \begin{cases} \alpha & \text{if } \alpha \leq 90° \\ 180° - \alpha & \text{if } \alpha > 90° \end{cases} \tag{5.9}$$

From the above equations it is obvious that, unlike in the θ-ϕ parametrization, α and β angles are not completely independent.

$$\boxed{\begin{array}{c} \alpha \text{ and } \beta \text{ angle parameters } \textit{can vary independently} \\ \text{if and only if } \phi \neq 0. \end{array}} \tag{5.10}$$

Furthermore, not all combinations of $\alpha \in [0, 180°]$ and $\beta \in [0, 180°]$ are acceptable/possible. The acceptable limits for α, when $\beta = \beta_o$, and the limits of β, when $\alpha = \alpha_o$, are independent of the value of the rotation of the x-y frame and may easily be derived from the limitations imposed by Eqs. (5.6) and (5.7). These limits are given below:

$$\text{if } \alpha = \alpha_o \text{ then}$$
$$\begin{cases} \overbrace{90° - \alpha_o}^{\triangleq \beta_{\min}} \leqslant \beta \leqslant \overbrace{90° + \alpha_o}^{\triangleq \beta_{\max}} & \text{for } \alpha_o \leqslant 90° \\ \underbrace{\alpha_o - 90°}_{\triangleq \beta_{\min}} \leqslant \beta \leqslant \underbrace{270° - \alpha_o}_{\triangleq \beta_{\max}} & \text{for } 90° < \alpha_o \leqslant 180° \end{cases} \tag{5.11}$$

$$\text{if } \beta = \beta_o \text{ then}$$
$$\begin{cases} \overbrace{90° - \beta_o}^{\triangleq \alpha_{\min}} \leqslant \alpha \leqslant \overbrace{90° + \beta_o}^{\triangleq \alpha_{\max}} & \text{for } \beta_o \leqslant 90° \\ \underbrace{\beta_o - 90°}_{\triangleq \alpha_{\min}} \leqslant \alpha \leqslant \underbrace{270° - \beta_o}_{\triangleq \alpha_{\max}} & \text{for } 90° < \beta_o \leqslant 180° \end{cases} \tag{5.12}$$

5.2 Manifold Vector in Terms of a Cone-Angle

Assume the array sensor positions are measured in half-wavelengths and represented by the matrix $\mathbf{r} = [\underline{r}_x, \underline{r}_y, \underline{r}_z]^T \in \mathcal{R}^{3 \times N}$ with respect to the Cartesian frame (x-y-z). If ($\check{\text{x}}$-$\check{\text{y}}$-z) is another Cartesian frame obtained by rotating the (x-y) axes on the (x,y) plane around the z-axis with a rotation angle Θ, then the sensor positions $\check{\mathbf{r}} \in \mathcal{R}^{3\times N}$ with respect to the Cartesian frame ($\check{x}$-$\check{y}$-z) can be expressed as follows:

$$\check{\mathbf{r}} = \begin{bmatrix} \cos\Theta & \sin\Theta & 0 \\ \cos(\Theta+90^\circ) & \sin(\Theta+90^\circ) & 0 \\ 0 & 0 & 1 \end{bmatrix} \overbrace{[\underline{r}_x, \underline{r}_y, \underline{r}_z]^T}^{=\mathbf{r}}$$

$$= [\underline{r}(\Theta), \underline{r}(\Theta+90^\circ), \underline{r}_z]^T \tag{5.13}$$

with

$$\underline{r}(\Theta) \triangleq \underline{r}_x \cos\Theta + \underline{r}_y \sin\Theta \tag{5.14}$$

$$\underline{r}(\Theta+90^\circ) \triangleq \underline{r}_x \cos(\Theta+90^\circ) + \underline{r}_y \sin(\Theta+90^\circ) \tag{5.15}$$

Based on the above, the wavenumber vector expressed in terms of the cone angles α and β, and for a rotation angle Θ, can be written as

$$\check{\underline{k}}(\alpha,\beta) = \pi\left[\cos\alpha, \cos\beta, \sqrt{1-\cos^2\alpha-\cos^2\beta}\right]^T \tag{5.16}$$

and thus, for isotropic sensors, the manifold vector of Eq. (1.22) can be expressed as a function of α and β parameters as follows:

$$\underline{a}(\alpha,\beta) = \exp(-j\check{\mathbf{r}}^T\check{\underline{k}}(\alpha,\beta)) \tag{5.17}$$

It should be emphasized here that Θ is a fixed scalar and *not* a direction of arrival parameter. Angles Θ and $\Theta+90^\circ$ simply define the lines of azimuth with respect to which α and β are respectively measured. Furthermore from Eqs. (5.14) and (5.15) it is easy to deduce that the roles of α and β can be interchanged by simply replacing Θ with $\Theta+90^\circ$.

5.3 Intrinsic Geometry of the Array Manifold Based on Cone-Angle Parametrization

Using the cone-angle directional parameters (α,β) instead of the generic parameters (p,q) of Chapter 3, the two-parameter array manifold $\mathcal{M}$ of Eq. (3.1) can be expressed as

$$\mathcal{M} = \{\underline{a}(\alpha,\beta) \in \mathfrak{C}^N,\ \forall(\alpha,\beta) : (\alpha,\beta) \in \Omega\} \tag{5.18}$$

where

$$\Omega = \{(\alpha,\beta) : \alpha \in [\alpha_{\min}, \alpha_{\max}) \quad \text{and} \quad \beta \in [\beta_{\min}, \beta_{\max})\}$$

with $\underline{\mathbf{a}}(\alpha, \beta)$ given by Eq. (5.17). As might be expected, the intrinsic shape of a manifold surface, as defined by its Gaussian curvature K_G, is invariant under any parameter transformation. For instance, using the general framework presented in Chapter 3 and replacing $p = \alpha$ and $q = \beta$, the Gaussian curvature of the manifold surface for a 3D grid array can be found to be

$$K_G = (\rho\pi)^{-2} \tag{5.19}$$

confirming (see Table 4.1) that the manifold is "developable" and that the shape of the manifold is independent of parametrization.

Next we will focus on the differential geometry properties of the manifold surfaces of planar arrays of omnidirectional sensors. Results for the 2D-grid arrays can easily be deduced from the following results and those for 3D-grid arrays can be analyzed with no difficulty.

Using Eq. (5.14) the manifold vector of Eq. (5.17) for a planar array is simplified to

$$\underline{\mathbf{a}}(\alpha, \beta) = \exp(-j\pi(\underline{r}(\Theta)\cos\alpha + \underline{r}(\Theta + 90°)\cos\beta)) \tag{5.20}$$

while its manifold surface $\mathcal{M}$ is represented by Eq. (5.18).

As usual, a study of intrinsic geometry begins with the specification of a metric tensor $\mathbb{G}$ and the evaluation of the first fundamental coefficients and the Christoffel symbols. It can be shown that the matrix $\mathbb{G}$ is given by

$$\mathbb{G} = \begin{bmatrix} g_{\alpha\alpha} & g_{\beta\alpha} \\ g_{\alpha\beta} & g_{\beta\beta} \end{bmatrix} = \begin{bmatrix} \|\dot{\underline{\mathbf{a}}}_\alpha\|^2 & \text{Re}(\dot{\underline{\mathbf{a}}}_\alpha^H \dot{\underline{\mathbf{a}}}_\beta) \\ \text{Re}(\dot{\underline{\mathbf{a}}}_\alpha^H \dot{\underline{\mathbf{a}}}_\beta) & \|\dot{\underline{\mathbf{a}}}_\beta\|^2 \end{bmatrix} \tag{5.21}$$

$$= \begin{bmatrix} \pi^2 \|\underline{r}(\Theta)\|^2 \sin^2\alpha, & \pi^2 \underline{r}(\Theta)^T \underline{r}(\Theta + 90°) \sin\alpha \sin\beta \\ \pi^2 \underline{r}(\Theta)^T \underline{r}(\Theta + 90°) \sin\alpha \sin\beta, & \pi^2 \|\underline{r}(\Theta + 90°)\|^2 \sin^2\beta \end{bmatrix}$$

where, as usual,

$$\dot{\underline{\mathbf{a}}}_\alpha = \frac{\partial \underline{\mathbf{a}}}{\partial \alpha} \tag{5.22}$$

and

$$\dot{\underline{\mathbf{a}}}_\beta = \frac{\partial \underline{\mathbf{a}}}{\partial \beta} \tag{5.23}$$

Whenever α or β equal 0 or π, the matrix $\mathbb{G}$ of the first fundamental coefficients becomes singular. At such singularities, one or both of the tangent vectors $\dot{\underline{\mathbf{a}}}_\alpha$ or $\dot{\underline{\mathbf{a}}}_\beta$ vanish and so the tangent plane is no longer defined. Fortunately, the requirement $\phi \neq 0$ (for the independence of α and β) established at the outset (see Eq. (5.10)) avoids these singularities. Also note that the cone parametrization avoids the singularity at the apex ($\phi = \frac{\pi}{2}$) which exists with θ-ϕ parametrization and hence is more suitable for the study of the manifold in that region.

The other variables required for the study of intrinsic geometry are the Christoffel symbols. From the definitions of Chapter 3, it can be shown that if

$$\mathbb{T} = [\underline{\dot{\mathbf{a}}}_\alpha, \ \underline{\dot{\mathbf{a}}}_\beta] \in \mathfrak{C}^{N \times 2} \tag{5.24}$$

then the Christoffel symbol matrices of first kind are

$$\left.\begin{aligned} \mathbf{\Gamma}_{1\alpha} &= \operatorname{Re}(\mathbb{T}^H \dot{\mathbb{T}}_\alpha) = \begin{bmatrix} \Gamma_{\alpha,\alpha\alpha} & \Gamma_{\alpha,\beta\alpha} \\ \Gamma_{\beta,\alpha\alpha} & \Gamma_{\beta,\beta\alpha} \end{bmatrix} \\ &= \cot\alpha \begin{bmatrix} g_{\alpha\alpha} & 0 \\ g_{\beta\alpha} & 0 \end{bmatrix} \\ \mathbf{\Gamma}_{1\beta} &= \operatorname{Re}(\mathbb{T}^H \dot{\mathbb{T}}_\beta) = \begin{bmatrix} \Gamma_{\alpha,\alpha\beta} & \Gamma_{\alpha,\beta\beta} \\ \Gamma_{\beta,\alpha\beta} & \Gamma_{\beta,\beta\beta} \end{bmatrix} \\ &= \cot\beta \begin{bmatrix} 0 & g_{\alpha\beta} \\ 0 & g_{\beta\beta} \end{bmatrix} \end{aligned}\right\} \tag{5.25}$$

while the matrices of second kind are given below:

$$\left.\begin{aligned} \mathbf{\Gamma}_{2\alpha} &= \mathbb{G}^{-1} \operatorname{Re}(\mathbb{T}^H \dot{\mathbb{T}}_\alpha) = \begin{bmatrix} \Gamma^\alpha_{\alpha\alpha} & \Gamma^\alpha_{\beta\alpha} \\ \Gamma^\beta_{\alpha\alpha} & \Gamma^\beta_{\beta\alpha} \end{bmatrix} \\ &= \cot\alpha \begin{bmatrix} 1 & 0 \\ 0 & 0 \end{bmatrix} \\ \mathbf{\Gamma}_{2\beta} &= \mathbb{G}^{-1} \operatorname{Re}(\mathbb{T}^H \dot{\mathbb{T}}_\beta) = \begin{bmatrix} \Gamma^\alpha_{\alpha\beta} & \Gamma^\alpha_{\beta\beta} \\ \Gamma^\beta_{\alpha\beta} & \Gamma^\beta_{\beta\beta} \end{bmatrix} \\ &= \cot\beta \begin{bmatrix} 0 & 0 \\ 0 & 1 \end{bmatrix} \end{aligned}\right\} \tag{5.26}$$

Note that the Christoffel matrices are quite sparse (even more so in the case of balanced-symmetric arrays where $g_{\alpha\beta} = g_{\beta\alpha} = 0$), which is a desirable feature since it simplifies the differential geometry considerably.

We have seen that the manifold surface of any planar array (see Table 4.2) has a zero Gaussian curvature. This is a property which is independent of the parametrization. Indeed following any of the Eqs. (3.31), (3.29) or (3.30), in conjunction with the Christoffel symbol matrices defined above (see Eqs. (5.25) and (5.26)), the Gaussian curvature of the array manifold can be found to be zero. Indeed, by using for instance Eq. (3.30), we have:

$$\begin{aligned}
K_{\text{gauss}} &= \frac{1}{\det[\mathbb{G}]}\begin{pmatrix} \frac{\partial}{\partial\alpha}\Gamma_{\alpha,\beta\beta} - \frac{\partial}{\partial\beta}\Gamma_{\alpha,\beta\alpha} \\ +\mathrm{col}_2\{\Gamma_{2\alpha}\}^T\mathrm{col}_1\{\Gamma_{1\beta}\} \\ -\mathrm{col}_2\{\Gamma_{2\beta}\}^T\mathrm{col}_1\{\Gamma_{1\alpha}\} \end{pmatrix} \\
&= \frac{1}{\det[\mathbb{G}]}\left(\cot\beta\frac{\partial g_{\alpha\beta}}{\partial\alpha} - 0 + 0 - g_{\beta\alpha}\cot\beta\cot\alpha\right) \\
&= \frac{1}{\det[\mathbb{G}]}(g_{\alpha\beta}\cot\alpha\cot\beta - g_{\beta\alpha}\cot\alpha\cot\beta) \qquad (5.27) \\
&= 0 \qquad (5.28)
\end{aligned}$$

The above result implies that the manifold is locally parabolic in $\mathfrak{C}^N$, in the same way that a circular cylinder, or cone, is locally parabolic in Euclidean space $\mathcal{R}^3$.

This also indicates that the array is developable which implies that the mapping of the manifold surface onto a real plane is distortionless.

Thus the array manifold, by virtue of its zero Gaussian curvature, is a developable surface and hence can be isometrically mapped onto any other surface of zero curvature, say a plane in $\mathcal{R}^2$.

5.4 Defining the Families of α- and β-parameter Curves

Based on the above discussion, and by keeping the frame-rotation angle Θ constant (with value Θ_0), the array manifold can also be described and analyzed by treating this surface as one of the following two families of curves:

(1) the family of α-parameter curves defined as

$$\{\mathcal{A}_{\alpha|\beta_o},\ \forall\beta_o : \beta_o \in \Omega_\beta\} \qquad (5.29)$$

where $\mathcal{A}_{\alpha|\beta_o}$ denotes the "α-curve" on the surface $\mathcal{M}$ corresponding to a constant value of $\beta = \beta_o$ and is represented as follows:

$$\mathcal{A}_{\alpha|\beta_o} = \{\underline{\mathbf{a}}(\alpha,\beta_o) \in \mathfrak{C}^N,\ \forall\alpha : \alpha \in \Omega_\alpha, \beta_o = \text{constant}\} \qquad (5.30)$$

(2) the family of β-parameter curves defined in a similar fashion as

$$\{\mathcal{A}_{\beta|\alpha_o},\ \forall\alpha_o : \alpha_o \in \Omega_\alpha\} \qquad (5.31)$$

where

$$\mathcal{A}_{\beta|\alpha_o} = \{\underline{\mathbf{a}}(\alpha_o,\beta) \in \mathfrak{C}^N,\ \forall\beta : \beta \in \Omega_\beta, \alpha_o = \text{constant}\} \qquad (5.32)$$

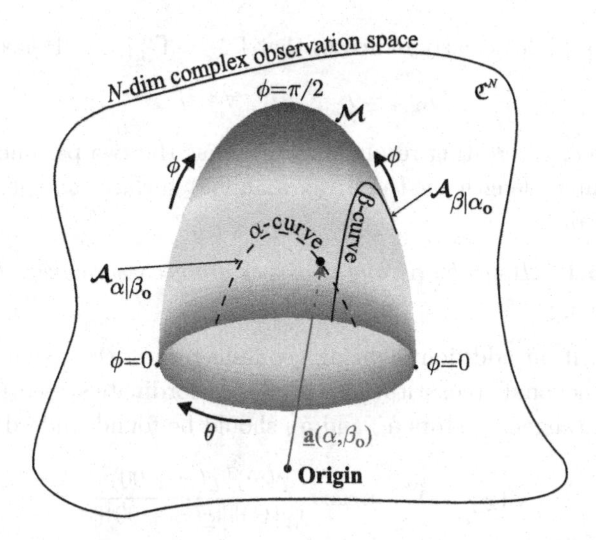

Fig. 5.4 An example of an α-curve and a β-curve lying on the array manifold $\mathcal{M}$.

In the previous equations $\Omega_\alpha \triangleq [\alpha_{\min} \text{ to } \alpha_{\max})$ and $\Omega_\beta \triangleq [\beta_{\min} \text{ to } \beta_{\max})$ with $\alpha_{\min}$, $\alpha_{\max}$, $\beta_{\min}$ and $\beta_{\max}$ given by Eqs. (5.12) and (5.11). Note that

$$\mathcal{M} = \left\{\mathcal{A}_{\alpha|\beta_o}, \ \forall \beta_o : \beta_o \in \Omega_\beta\right\} = \left\{\mathcal{A}_{\beta|\alpha_o}, \ \forall \alpha_o : \alpha_o \in \Omega_\alpha\right\} \tag{5.33}$$

Fig. 5.4 shows an illustrative representation of one "α-curve" and one "β-curve" lying on the array manifold surface $\mathcal{M}$.

The differential geometry of the α- and β-curves on the array manifold surface are next investigated. Naturally, due to the symmetry of Eq. (5.20), the parameter curves are expected to exhibit similar characteristics.

5.5 Properties of α- and β-parameter Curves

5.5.1 *Geodecity*

Following the definitions of Chapter 3, the geodesic curvatures of the α- and β-curves (see Eqs. (3.42) and (3.43)) are given by

$$\begin{cases} \kappa_{g,\alpha} = \sqrt{\det[\mathbb{G}]}\,\dfrac{\Gamma^{\beta}_{\alpha\alpha}}{(g_{\alpha\alpha})^{3/2}} \\ \kappa_{g,\beta} = -\sqrt{\det[\mathbb{G}]}\,\dfrac{\Gamma^{\alpha}_{\beta\beta}}{(g_{\beta\beta})^{3/2}} \end{cases} \tag{5.34}$$

However, Eq. (5.26) clearly indicates that $\Gamma^{\beta}_{\alpha\alpha} = \Gamma^{\alpha}_{\beta\beta} = 0$. Hence

$$\kappa_{g,\alpha} = \kappa_{g,\beta} = 0 \quad \forall \alpha, \beta \tag{5.35}$$

which is a very interesting result, implying that the two parameter curves are of minimum length on the array manifold surface, and leads to the following theorem.

Theorem 5.1 *Both the α- and β-curves are geodesic curves for all values of Θ.*

To check if, in addition to being geodesic curves, the α- and β-curves are also "orthogonal" (constituting a geodesic coordinate system), the angle between the tangent vectors $\underline{\dot{\mathbf{a}}}_\alpha$ and $\underline{\dot{\mathbf{a}}}_\beta$ should be found. Indeed

$$\begin{aligned}\measuredangle(\underline{\dot{\mathbf{a}}}_\alpha, \underline{\dot{\mathbf{a}}}_\beta) &= \arccos \frac{\underline{r}(\Theta)^T \underline{r}(\Theta + 90)}{\|\underline{r}(\Theta)\| \|\underline{r}(\Theta + 90)\|} \\ &= \measuredangle(\underline{r}(\Theta)^T \underline{r}(\Theta + 90)) \end{aligned} \tag{5.36}$$

and is, hence, constant. If $\underline{r}_x \perp \underline{r}_y$ then

$$\underline{r}(\Theta)^T \underline{r}(\Theta + 90^\circ) = \tfrac{1}{2}(\|\underline{r}_y\|^2 - \|\underline{r}_x\|^2) \sin(2\Theta)$$

and the tangent vectors $\underline{\dot{\mathbf{a}}}_\alpha$ and $\underline{\dot{\mathbf{a}}}_\beta$ (and therefore α- and β-curves) are strictly orthogonal and constitute geodesic coordinates if $\Theta = 0$, or $\|\underline{r}_x\| = \|\underline{r}_y\|$. Thus, in summary,

$$\left.\begin{array}{l} \text{(i)}\ \ (\alpha\text{-curves}) \text{ and } (\beta\text{-curves}) : \text{geodesic curves} \\ \qquad \text{i.e. } \kappa_{g,\alpha} = 0 = \kappa_{g,\beta}\ \forall\Theta \\ \text{(ii)}\ \text{if} \left\{ \begin{array}{l} \Theta = 0 \text{ and } \underline{r}_x \perp \underline{r}_y, \text{ or} \\ \|\underline{r}_x\| = \|\underline{r}_y\| \text{ and } \underline{r}_x \perp \underline{r}_y \end{array} \right. \\ \qquad \text{then } (\alpha\text{-curves}) \perp (\beta\text{-curves}) \Longrightarrow \textit{geodesic} \text{ coordinates} \end{array} \right\} \tag{5.37}$$

It is a simple matter to confirm that for 2D-grid (balanced-symmetric[1]) arrays

$$\underline{r}(\Theta)^T \underline{r}\,(\Theta + 90^\circ) = 0, \quad \forall\Theta \tag{5.38}$$

and so the tangent vectors $\underline{\dot{\mathbf{a}}}_\alpha$ and $\underline{\dot{\mathbf{a}}}_\beta$ are strictly orthogonal. For such arrays matrix $\mathbb{G}$ is diagonal and the hyperhelical α-β curves form an orthogonal coordinate system $\forall\Theta$.

---

[1]• balanced-symmetric arrays: $\|\underline{r}_x\| = \|\underline{r}_y\|$ and $\underline{r}_x \perp \underline{r}_y$
• unbalanced-symmetric arrays: $\|\underline{r}_x\| \neq \|\underline{r}_y\|$ and $\underline{r}_x \perp \underline{r}_y$

5.5.2 *Length of Parameter Curves*

As was stated before, the most basic features of a curve $\underline{a}(p)$ are its arc length s and rate of change of arc length $\dot{s}(p)$

$$\dot{s}(p) = \frac{ds}{dp} = |\dot{\underline{a}}(p)| \tag{5.39}$$

In the case of the α- and β-parameter curves (assuming $\phi \neq 0$) for a given rotation Θ_o, the arc length of an α-curve $\mathcal{A}_{\alpha|\beta_o}$ and of a β-curve $\mathcal{A}_{\beta|\alpha_o}$ are respectively

$$s(\alpha) = \pi\|\underline{r}(\Theta_o)\|(1 - \cos\alpha) \tag{5.40}$$

$$s(\beta) = \pi\|\underline{r}(\Theta_o + 90°)\|(1 - \cos\beta) \tag{5.41}$$

while their rate of change of arc length can be expressed as

$$\dot{s}(\alpha) = \pi\|\underline{r}(\Theta_o)\| \sin\alpha; \qquad \forall\alpha \in \Omega_{\alpha|\beta_o} = [\alpha_{\min}, \alpha_{\max}] \tag{5.42}$$

$$\dot{s}(\beta) = \pi\|\underline{r}(\Theta_o + 90°)\| \sin\beta; \quad \forall\beta \in \Omega_{\beta|\alpha_o} = [\beta_{\min}, \beta_{\max}] \tag{5.43}$$

Then, by integrating Eq. (5.42) over its limits $\alpha_{\min}$ and $\alpha_{\max}$, the length l_m of an α-curve can be found as

$$\mathcal{A}_{\alpha|\beta_o} : l_m = 2\pi\|\underline{r}(\Theta_o)\| \sin\beta_0 \tag{5.44}$$

which indicates that α-curves, although identical in shape, have different lengths, with the α-curve associated with $\beta_o = 90°$ having the maximum length and their lengths gradually reducing as β_o increases from 90° to 180° or decreases from 90° to 0°.

In a similar fashion the length of a β-curve, is as follows

$$\mathcal{A}_{\beta|\alpha_o} : l_m = 2\pi\|\underline{r}(\Theta_o + 90°)\| \sin\alpha_o \tag{5.45}$$

Clearly the α- (or β-) curve degenerates into a single point (i.e. a curve of zero length) whenever β (or α) approaches either 0 or π radians.

5.5.3 *Shape of α- and β-curves*

The shapes of the parameter curves can be identified through evaluation of their curvatures. It can be found that the curvatures of the α- and β-parameter curves are constant and can be found from the recursive Eq. (2.39) with the difference that $\underline{r}(\Theta)$ and $\underline{r}(\Theta + \frac{\pi}{2})$, respectively, have taken on the role of the sensor location vector $\underline{r}$ of the linear array. This is apparent from Eq. (2.39) which indicates that curvatures are independent of α and β respectively and are only functions of the array geometry projected along the lines of azimuth at Θ and $\Theta + \frac{\pi}{2}$. Consequently, the curvatures and hence the shapes of the α, β-curves depend strongly on Θ, reminiscent of the drastic variations in array performance which can be observed along different lines of azimuth. Thus, Theorem 2.1 is the centrepiece of the investigation of the α/β-curves and, as will be demonstrated next, has some far-reaching implications.

Corollary 5.1 *The α- and β-curves are* both *hyperhelical.*

For instance for an α-curve, of constant β ($\beta = \beta_o$) and constant Θ ($\Theta = \Theta_o$), Eq. (5.20) becomes identical to Eq. (2.27) with

$$\underline{r} = \underline{r}(\Theta_o), \qquad p = \alpha, \qquad \underline{v} = \underline{r}(\Theta_o + 90°)\cos\beta_o = \text{constant} \tag{5.46}$$

From the above it is clear that all α-curves have the same ELA $\underline{r}(\Theta_o)$ but different "visible" areas specified by the minimum and maximum permissible values of α, i.e. $\Omega_{\alpha|\beta_o} = [\alpha_{\min}, \alpha_{\max}]$. The same is true for the β-curves but with $\underline{r}(\Theta_o + 90°)$ as their ELA with $\Omega_{\beta|\alpha_o} = [\beta_{\min}, \beta_{\max}]$.

$$\boxed{\begin{array}{ll}\text{ELA of the family } \{\mathcal{A}_{\alpha|\beta_o} \forall\beta_o : \beta_o \in \Omega_\beta\}: & \underline{r}(\Theta_o) \\ \text{ELA of the family } \{\mathcal{A}_{\beta|\alpha_o} \forall\alpha_o : \alpha_o \in \Omega_\alpha\}: & \underline{r}(\Theta_o + 90°)\end{array}} \tag{5.47}$$

Figure 5.5 provides an illustrative picture of an α-curve highlighting its hyperhelical shape and the fact that its length increases for values of β_o approaching 90°.

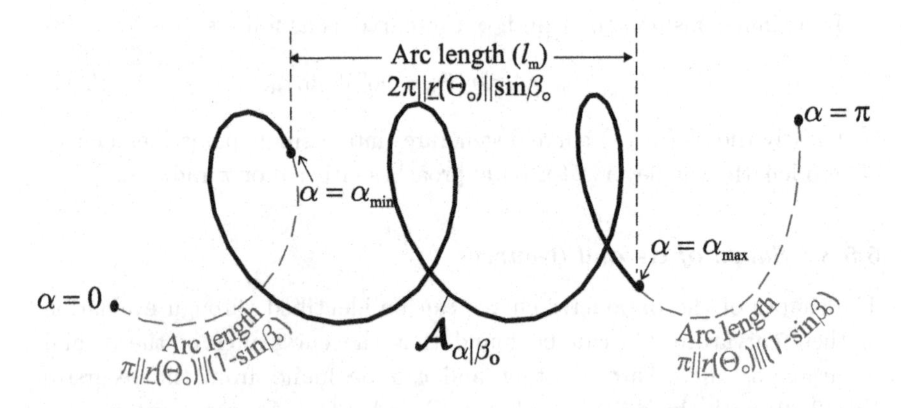

Fig. 5.5 An illustrative diagram of the length of the hyperhelical α-curve $\mathcal{A}_{\alpha|\beta_o}$.

Corollary 5.2 *For a given frame rotation angle Θ_o, all members of the family of α- (or β-) curves of a planar array of N sensors are identical (congruent) in shape apart from their length and position in complex N-dimensional space.*

The above result means that the whole of the manifold surface may be fully covered by the simple translation of either of the two individual curves with displacement values given by $\underline{v} = \underline{r}(\Theta_o + 90°)\cos\beta_0$ for α-curves and $\underline{v} = \underline{r}(\Theta_o)\cos\alpha_0$ for β-curves. Note that α-curves are the same as β-curves for rotation $\Theta_o + 90°$.

Corollary 5.3 *The α- and β-curves are identical to the θ-parameter manifold of a linear array of isotropic sensors with locations $\underline{r}(\Theta)$ and $\underline{r}(\Theta + \frac{\pi}{2})$ respectively.*

Figure 5.6 illustrates the α-β parametrization of the array manifold in the special case of $\Theta = 0$. Notice that for $\Theta = 0$, cone-angles are measured with respect to the x and y axes and consequently, according to Corollary 5.3, the α- and β-curves are identical to the θ-parameter manifolds of linear arrays with sensor locations $\underline{r}(0) = \underline{r}_x$ and $\underline{r}(\frac{\pi}{2}) = \underline{r}_y$ respectively.

Although the hyperhelical α (or β) family of curves shown above covers the whole of the manifold surface and is sufficient to fully define the manifold shape, it is not unique. In fact, to understand the behavior of the array for all possible directions of arrival, it is necessary to consider the hyperhelical coordinates corresponding to all values of Θ from 0 to $\frac{\pi}{2}$.

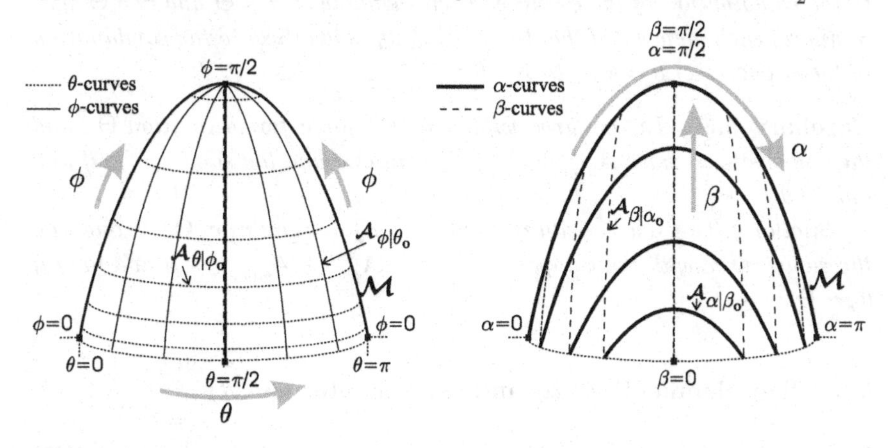

Fig. 5.6 (θ, ϕ) and (α, β) parametrizations of the same manifold surface $\mathcal{M}$ with the frame rotation for the cone-angles $\Theta_o = 0$.

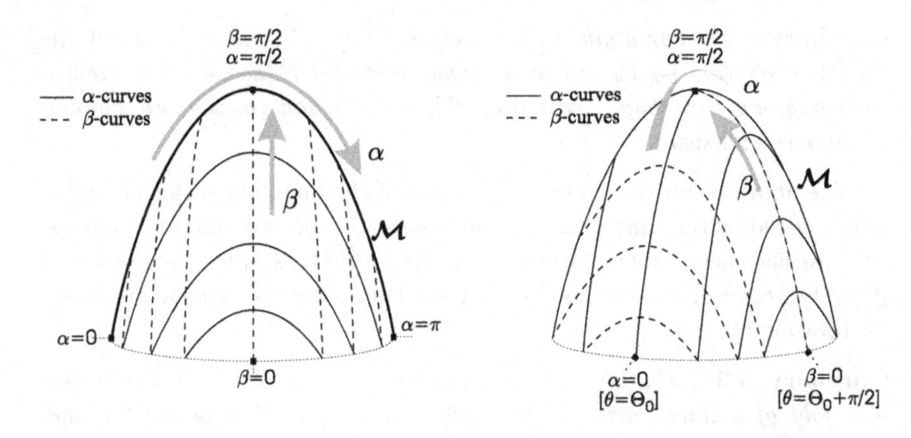

Fig. 5.7 Families of α and β curves for a frame rotation $\Theta_o = 0°$. and for $\Theta_o < 90°$.

The above concept is more easily appreciated by observing Fig. 5.7 which shows the effect of a non-zero value of Θ. It should be stressed that, despite the appearance of Fig. 5.7, the use of various values of Θ is *not* equivalent to a trivial rotation of the parameter curves, since their differential geometry is a function of Θ (Corollary 5.3).

Figures 5.6 and 5.7 also reveal a clear relationship between the α (or β) curves and the hyperhelical ϕ-curves, which will be stated as the final two corollaries:

Corollary 5.4 *Each member of the α-curve family is identical in shape to the combination of the ϕ-curves corresponding to $\theta = \Theta$ and $\theta = \Theta + \pi$. Similarly each member of the β-curve family is identical to the combination of the ϕ-curves at $\theta = \Theta + \frac{\pi}{2}$ and $\theta = \Theta + \frac{3\pi}{2}$.*

Corollary 5.5 *The α-curve with $\beta = 90°$ for a frame rotation Θ_o and the composite ϕ curve $\mathcal{A}_{\phi|\theta_o} + \mathcal{A}_{\phi|(\theta_o+\pi)}$ are one and the same curve if and only if $\theta_o = \Theta_o$.*

Similarly the β-curve with $\alpha = 90°$ for a frame rotation Θ_o is one and the same curve with the composite ϕ curve $\mathcal{A}_{\phi|\theta_o} + \mathcal{A}_{\phi|(\theta_o+\pi)}$ if and only if $\theta_o = \Theta_o + \frac{\pi}{2}$.

5.6 "Development" of α- and β-parameter Curves

The most significant implication of the property that the array manifold $\mathcal{M} \in \mathfrak{C}^N$ of a planar array has a zero Gaussian curvature, is that $\mathcal{M}$ is a developable surface and hence can be isometrically mapped on to any other surface of zero curvature, say a plane in $\mathcal{R}^2$. The result of such a mapping would be a consistent two-dimensional representation (development) of the manifold in real space. A development may be readily evaluated by noting that if $\mathcal{A}_{\alpha|\beta_o}, \mathcal{A}_{\beta|\alpha_o} \in \mathfrak{C}^N$ are arbitrary α and β curves on the manifold surface and $\mathcal{A}^{(d)}_{\alpha|\beta_o}, \mathcal{A}^{(d)}_{\beta|\alpha_o} \in \mathcal{R}^2$ are their respective images on the development, then since an isometric mapping is by necessity both geodesic and conformal:

- Length of $\mathcal{A}^{(d)}_{\alpha|\beta_o} \in \mathcal{R}^2$ = Length of $\mathcal{A}_{\alpha|\beta_o} \in \mathfrak{C}^N$,
- Principal curvature of $\mathcal{A}^{(d)}_{\alpha|\beta_o} \in \mathcal{R}^2$ = Geodesic curvature of $\mathcal{A}_{\alpha|\beta_o} \in \mathfrak{C}^N$,
- $\measuredangle(\mathcal{A}^{(d)}_{\alpha|\beta_o}, \mathcal{A}^{(d)}_{\beta|\alpha_o})$ in $\mathcal{R}^2$ = $\measuredangle(\mathcal{A}_{\alpha|\beta_o}, \mathcal{A}_{\beta|\alpha_o})$ in $\mathfrak{C}^N$.

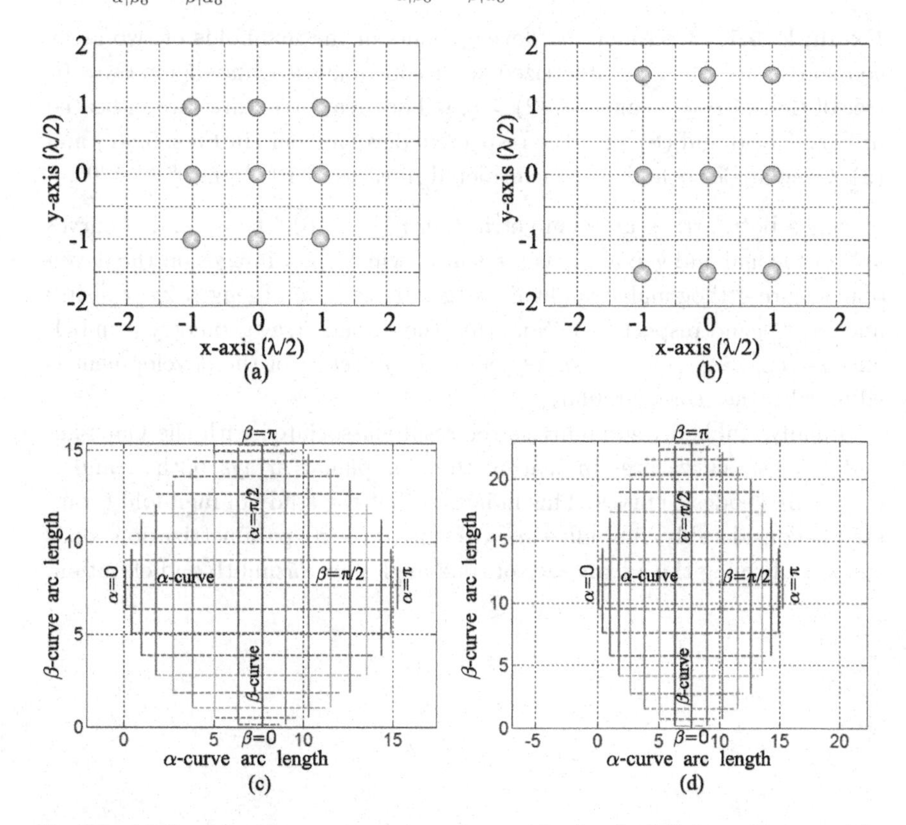

Fig. 5.8 (a) Square-grid array structure, $\underline{r}_y \perp \underline{r}_x$ and $\|\underline{r}_y\| = \|\underline{r}_x\|$ (balanced-symmetric). (b) Rectangular-grid array structure $\underline{r}_y \perp \underline{r}_x$ and $\underline{r}_y = 1.5\ \underline{r}_x$ (unbalanced-symmetric). (c) and (d) Developments of the manifolds of arrays in Figs. (a) and (b) respectively when the manifolds are parametrized by cone-angles α and β with frame rotation $\Theta = 0°$.

Table 5.1 Cone-parametrization manifold parameters for 3D-grid and planar arrays.

Parameter	3D	Planar
K_G (Eq. (3.31))	$\frac{1}{\rho^2\pi^2}$	0
$\kappa_{g,\alpha}$	$\frac{\cot\beta}{\rho\pi}$	0
$\kappa_{g,\beta}$	$\frac{\cot\alpha}{\rho\pi}$	0

The above features may be demonstrated by the following example.

Example 5.1 Consider the developments of the manifolds of two array configurations when parametrized with cone-angles α and β for $\Theta = 0°$ (recall that $\underline{r}(0) = \underline{r}_x$ and $\underline{r}(\pi/2) = \underline{r}_y$). The array structures are indicated in Figs. 5.8(a) and (b) with the respective developments in Figs. 5.8(c) and (d). Cone-angles α and β are considered in steps of 10° from 0° to 180°.

Since both arrays are symmetric (i.e. $\underline{r}_x^T\underline{r}_y = 0$), the α- and β-curves are orthogonal and geodesic, as a result of which their images on the development are orthogonal straight lines (geodesics) and of length $2\pi\|\underline{r}_x\|\sin\beta$ and $2\pi\|\underline{r}_y\|\sin\alpha$ respectively. Note that the second array geometry is unbalanced ($\|\underline{r}_y\| > \|\underline{r}_x\|$) and consequently the *envelope* of the development is elliptical rather than circular.

Finally, Table 5.1 summarizes the results associated with the Gaussian and geodesic curvatures indicating that for planar arrays both α and β curves are geodesic curves. This indicates that the study of manifold geometry is considerably simplified as a result of (α, β) parametrization, once again confirming the advantages of cone-angles over azimuth and elevation.

Chapter 6

Array Ambiguities

In this chapter ambiguous sets of parameter values (e.g. directions) that exist in any hyperhelical curve $\mathcal{A}$ embedded in an N dimensional complex space, are modelled, identified and calculated. The curve $\mathcal{A}$, which is defined as follows

$$\mathcal{A} = \{\underline{\mathbf{a}}(p) \in \mathfrak{C}^N, \ \forall p : p \in \Omega_p\} \tag{6.1}$$

with

$$\underline{\mathbf{a}}(p) = \exp(-j(\pi \underline{r} \cos p + \underline{v})) \tag{6.2}$$

could be the manifold of a linear array or a p-curve (with $p = \phi, \alpha$ or β) lying on the manifold surface of a planar array (using the concept of the ELA). In these cases the manifold vector $\underline{\mathbf{a}}(p)$ represents the complex array response to a unity power signal impinging on the array from direction p and is a function of the locations $\underline{r} \in \mathcal{R}^N$ of the array elements, defined as

$$\text{sensor locations} : \underline{r} \triangleq \underline{r}(\Psi_0) = \underline{r}_x \cos \Psi_0 + \underline{r}_y \sin \Psi_0 \tag{6.3}$$

The parameter Ψ_0, shown in Fig. 6.1 and used in Eq. (6.3), is defined as

$$\boxed{\Psi_0 \triangleq \begin{cases} 0^\circ & \text{for any linear array on the } x\text{-axis} \\ \theta_o & \text{for one } \phi\text{-curve } \mathcal{A}_{\phi|\theta_o} \\ \Theta_o & \text{for all } \alpha\text{-curves} \\ \Theta_o + 90^\circ & \text{for all } \beta\text{-curves} \end{cases}} \tag{6.4}$$

where θ_o is the constant azimuth (ϕ-curves) and Θ_o is the frame rotation angle (α-curves). Table 6.1 provides the main parameters of Eq. (6.2) for the various hyperhelical array manifold curves.

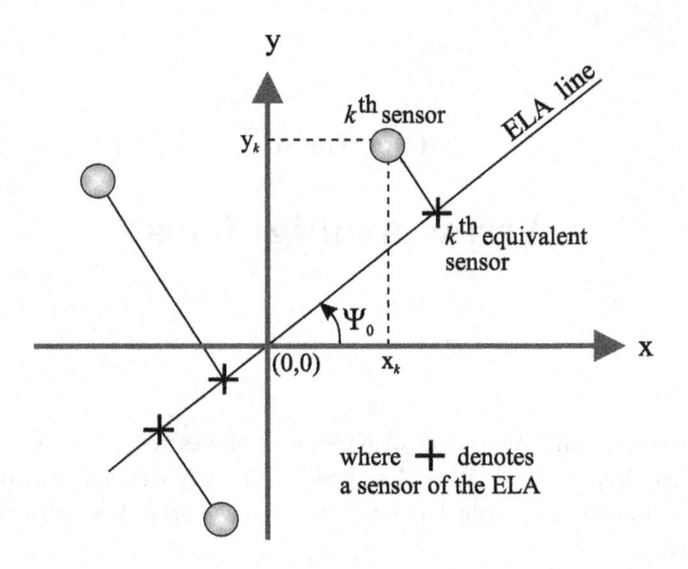

Fig. 6.1 Equivalent linear array $\underline{r}(\Psi_0)$.

Table 6.1 Array manifold vector parameters (Eq. (6.2)).

Parameter	Linear Arrays	Planar Arrays		
p (generic)	θ	ϕ	α	β
$\Omega_p \in \mathcal{R}^1$	$[0,\pi)$	$[0,\pi/2)$	$[0,\pi)$	$[0,\pi)$
$\underline{r}$	$\underline{r}_x$	$\underline{r}(\theta_o)$	$\underline{r}(\Theta_o)$	$\underline{r}(\Theta_o+90°)$
$\underline{h}(p)=\cos(p)$	$\cos(\theta)$	$\cos(\phi)$	$\cos(\alpha)$	$\cos(\beta)$
$\underline{v}$	0	0	$\underline{r}(\Theta_o+90°)\cos(\beta_o)$	$\underline{r}(\Theta_o)\cos(\alpha_o)$
curve	$\mathcal{A}$	$\mathcal{A}_{\phi\mid\theta_o}$	$\mathcal{A}_{\alpha\mid\beta_o}$	$\mathcal{A}_{\beta\mid\alpha_o}$

6.1 Classification of Ambiguities

The array manifold $\mathcal{A}$ of Eq. (6.1) is essentially a mapping from the parameter space $\Omega_p \in \mathcal{R}^1$ to the complex N-dimensional space $\mathfrak{C}^N$

$$p \in \Omega_p \overset{\textbf{array geometry}}{\longmapsto} \underline{\mathbf{a}}(p) \in \mathfrak{C}^N \tag{6.5}$$

When finding the directions-of-arrival (DOA's) of signals using array sensor measurements, it is important to be certain that the problem has a unique solution. This is guaranteed if the above mapping is *one-to-one*.

However, in this study we will consider the problem which arises when the mapping of Eq. (6.5) is not *one-to-one*. In this case the array cannot distinguish between two (or more) different signal environments. For instance two different sets of signals impinging on the array can provide identical responses at the array output, i.e. the same measurements. In such a case any estimation algorithm may be unable to distinguish the true parameter values (e.g. directions) from the false ones. This is termed as the ambiguity problem.

Definition 6.1 — Array Ambiguity: It is the inability of an array of a given geometry to distinguish a set of parameter values

$$(p_1, p_2, \ldots, p_M) \quad \text{with} \quad p_i \neq p_j \in \Omega_p \quad \text{for } i \neq j \tag{6.6}$$

from at least one of its subsets (or one subset from another).

Array ambiguities can be classified as "trivial" and "non-trivial." We will start with trivial ambiguities.

Definition 6.2 — Trivial Ambiguities: An array ambiguity is said to be classified as 'trivial' if there exist at least one pair $p_1, p_2 \in \Omega_p$ with $p_1 \neq p_2$ such that

$$\boxed{\begin{array}{lcl} p_1 & \overset{\textbf{array geometry}}{\longmapsto} & \underline{\mathbf{a}}(p_1) \in \mathfrak{C}^N \\ p_2 & \overset{\textbf{array geometry}}{\longmapsto} & \underline{\mathbf{a}}(p_2) \in \mathfrak{C}^N \end{array}} \tag{6.7}$$

with

$$\underline{\mathbf{a}}(p_1) = k\underline{\mathbf{a}}(p_2) \tag{6.8}$$

and

$$k = \text{scalar}$$

In this case the two parameter values p_1, p_2 are indistinguishable by the array.

Trivial ambiguities are the simplest type of ambiguity and can be easily identified/detected. For instance it is impossible to distinguish whether a signal is impinging on a linear array from direction-of-arrival p_1 or from the direction $p_2 = 360° - p_1$, i.e. the mirror image of p_1 with respect to 180°. In this case

$$\underline{\mathbf{a}}(p_1) = \underline{\mathbf{a}}(p_2) \tag{6.9}$$

and for this reason, in order to avoid this ambiguity problem, the parameter space in the case of linear arrays is confined to $\Omega_p = [0°, 180°)$.

However, Eq. (6.9) is a special case of Eq. (6.8) for $k = 1$. Eq. (6.8) indicates that there exists at least one manifold vector $\underline{\mathbf{a}}(p_2)$ which is colinear to another manifold vector $\underline{\mathbf{a}}(p_1)$. Then the two waves with bearings p_1, p_2 are indistinguishable by the array even if $p_1, p_2 \in \Omega_p = [0°, 180°)$.

Non-trivial ambiguities are a more complicated type of ambiguity arising when a manifold vector can be written as a linear combination of two or more different manifold vectors.

Definition 6.3 — Non-Trivial Ambiguities: An array ambiguity is said to be classified as "non-trivial" if there exist at least a set of values $p_1, p_2, \ldots, p_M \in \Omega_p$ with $M > 2$ and $p_i \neq p_j$ for $i \neq j$ such that

$$p_i \in \Omega_p \overset{\textbf{array geometry}}{\longmapsto} \underline{\mathbf{a}}(p_i) \in \mathfrak{C}^N; \quad \forall i = 1, 2, \ldots, M \tag{6.10}$$

with

$$\underline{\mathbf{a}}(p_M) = c_1\underline{\mathbf{a}}(p_1) + c_2\underline{\mathbf{a}}(p_2) + \cdots + c_{M-1}\underline{\mathbf{a}}(p_{M-1}) \tag{6.11}$$

Then this set $(p_1, p_2, \ldots, p_M)$ and all its subsets of $M - 1$ elements are indistinguishable by the array.

In such a case the array will have identical responses for the sets of bearings $(p_1, p_2, \ldots, p_{M-1})$, $(p_2, p_3, \ldots, p_M)$, $(p_1, p_3, p_4, \ldots, p_M)$, etc.

Fig. 6.2 illustrates a non-trivial ambiguous situation where the manifold vector $\underline{\mathbf{a}}(p_3)$ belongs to the subspace $\mathcal{L}[\mathbb{A}]$ where $\mathbb{A} = [\underline{\mathbf{a}}(p_1), \underline{\mathbf{a}}(p_2)]$.

This type of ambiguity is much more difficult to identify since, unlike a trivial ambiguity, it cannot be detected by a simple search of the manifold. For example consider the non-uniform linear array of four sensors with sensor locations (measured in half wavelengths) given by

$$\underline{r} = [-2, -1, 1, 2]^T \tag{6.12}$$

One non-trivial ambiguity of this array is related to the set of directions

$$0°, 60°, 90°, 120°$$

as the array is unable to distinguish between the following signal environments:

$$(0°, 60°, 90°), (0°, 60°, 120°), (60°, 90°, 120°), (0°, 60°, 90°, 120°)$$

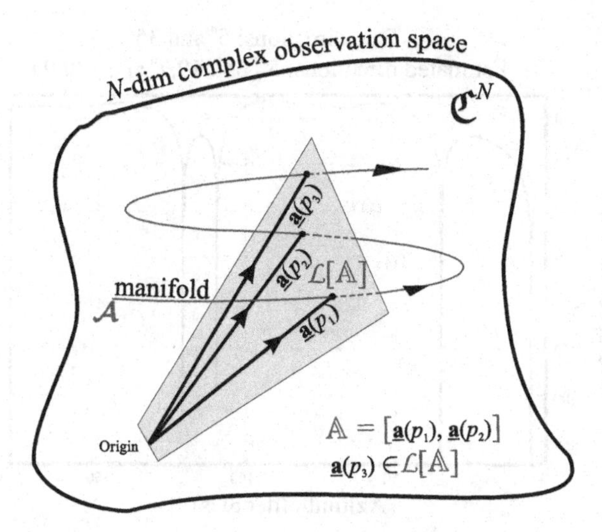

Fig. 6.2 Illustrative representation of array ambiguity (the vectors $\underline{\mathbf{a}}(p_1), \underline{\mathbf{a}}(p_2)$ and $\underline{\mathbf{a}}(p_3)$ are linearly dependent).

For instance, for this array, the above four different signal environments provide identical MUSIC spectra. Note that for this specific array,

$$\mathbf{rank}\,([\underline{\mathbf{a}}(0^\circ),\ \underline{\mathbf{a}}(60^\circ),\ \underline{\mathbf{a}}(90^\circ),\ \underline{\mathbf{a}}(120^\circ)]) = 3 \tag{6.13}$$

and hence these four manifold vectors are linearly dependent.

In general we can say that every array suffers from ambiguities in some way or another as a result of the array geometry and thus different array geometries have different sets of ambiguous directions. This implies that an array which is unambiguous for a given set of directions might become ambiguous for the same set if we change its sensor locations even slightly to form a new array geometry. A typical example can be seen using the following linear array,

$$\underline{r} = [-3, -1.5, 0, \underset{\longleftrightarrow}{1.0}, 3]^T \tag{6.14}$$

operating in the presence of two sources from unambiguous directions of 5° and 35°. If the fourth sensor located at +1 half-wavelengths is now moved to the location +1.5 half-wavelengths, so that the uniform linear array

$$\underline{r} = [-3, -1.5, 0, \underset{\longleftrightarrow}{1.5}, 3]^T \tag{6.15}$$

is obtained, then these two sources at 5° and 35° become "ambiguous" as these two directions are now a subset of the "non-trivial" type ambiguous set of directions 5°, 35°, 109.70°, 120.94°. This can also be seen, for instance,

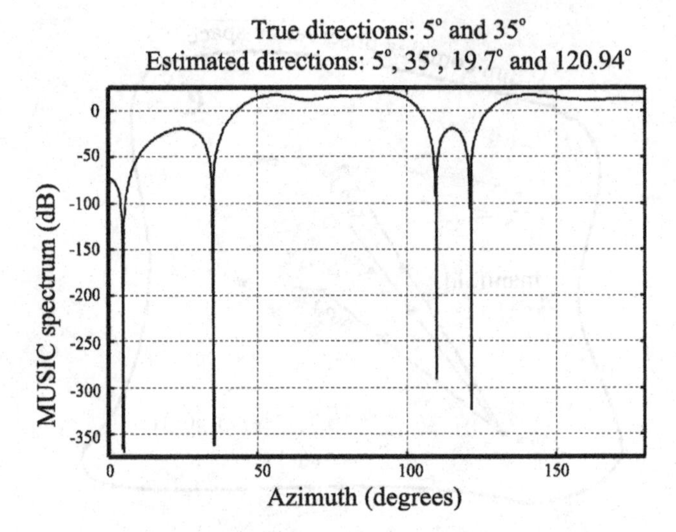

Fig. 6.3 MUSIC spectrum in an "ambiguous" situation.

from Fig. 6.3 where in the MUSIC spectrum, four nulls appear at directions 5°, 35°, 109.70°, 120.94° rather than two at 5° and 35°.

Note that in this study the array centroid will be taken as the reference point, implying that $\mathbf{sum}(\underline{r}) = \underline{r}^T \underline{1}_N = 0$. However, it can be proved that the ambiguities are independent of the choice of the array reference point.

6.2 The Concept of an Ambiguous Generator Set

Before continuing it is necessary to present the following definitions which will be extensively used.

Definition 6.4 — Ambiguous Set and Rank of Ambiguity

(1) Ambiguous Set:
An ordered set of arc lengths $\underline{s} = [s_1, s_2, \ldots, s_c]^T$, is said to be an *ambiguous set of arc lengths*

if $c \leq N$ and the matrix $\mathbb{A}(\underline{s}) \in \mathfrak{C}^{N \times c}$ with columns the manifold vectors $\underline{\mathbf{a}}(s_1)$, $\underline{\mathbf{a}}(s_2)$, ..., $\underline{\mathbf{a}}(s_c)$ has rank less than c, i.e. $\mathbf{rank}(\mathbb{A}(\underline{s})) < c$

or

if $c > N$ and all subsets $\underline{s}_i$ of $\underline{s}$ which contain exactly N elements are themselves ambiguous sets of arc lengths,

$$\text{i.e.} \quad \mathbf{rank}(\mathbb{A}(\underline{s}_i)) < N, \quad \forall i \tag{6.16}$$

(2) Rank of Ambiguity:
For an ambiguous set of arc lengths $\underline{s} = [s_1, s_2, \ldots, s_c]^T$ its rank of ambiguity is defined as the integer ρ_a

$$\text{where } \rho_a = \begin{cases} \mathbf{rank}(\mathbb{A}(\underline{s})) & \text{if } c \leqslant N \\ \min\limits_{\forall i} \{\mathbf{rank}\,(\mathbb{A}(\underline{s}_i))\} & \text{if } c > N \end{cases} \tag{6.17}$$

with $\underline{s}_i \in \mathcal{R}^N, \forall i$.

Note that the previously introduced definitions can be directly applied to sets of directional parameters by simply substituting (using Eq. (2.28)) the arc length s with p.

The following theorem, which is a very important result, essentially states that if all the elements of an ambiguous set of arc lengths are rotated on the array manifold by the same value, then the resulting set is also an ambiguous set of arc lengths.

Theorem 6.1 — ***Arc length Rotation:*** *For a linear array of N sensors, if*

$$\underline{s} = [s_1, s_2, \ldots, s_c]^T \tag{6.18}$$

is an ambiguous set of arc lengths with rank of ambiguity ρ_a, then any set

$$\begin{aligned} \underline{\check{s}} &= [s_1 + \Delta s, s_2 + \Delta s, \ldots, s_c + \Delta s]^T \\ &= \underline{s} + \Delta s.\underline{1}_c, \end{aligned} \tag{6.19}$$

with $s_1 + \Delta s \geq 0$ and $s_c + \Delta s \leq l_m$, is also an ambiguous set of arc lengths with the same rank of ambiguity ρ_a.

The proof of this theorem can be found in Appendix 6.7.1 while its essential features are illustrated in the following example.

Example 6.1 Consider a linear array having sensor locations given, in half-wavelengths, by the following vector:

$$\underline{r} = [-3, -1, -0.5, 2, 2.5]^T$$

and having a manifold length $l_m = 2\pi\|\underline{r}\| = 28.4483$. It can easily be seen that the set

$$\underline{p} = [0°, 53.13°, 78.46°, 101.54°, 126.87°]^T$$

is an ambiguous set of directions. Using Eq. (2.28) the corresponding ambiguous set of arc lengths is computed to be

$$\underline{s} = [0, 5.6897, 11.3793, 17.0690, 22.7857]^T$$

If this set $\underline{s}$ is rotated by $\Delta s = 2$, then a new set of arc lengths

$$\underline{\check{s}} = [2, 7.6897, 13.3793, 19.0690, 24.7857]^T$$

is obtained, which corresponds to the set of directions

$$\underline{\check{p}} = [30.75°, 62.65°, 86.59°, 109.91°, 137.78°]^T$$

which is also an ambiguous set. This is also true for any Δs so long as the last element of $\underline{\check{s}}$ is smaller than, or equal to, the manifold length l_{m}. Note that the rotation should be carried out in the "arc length" domain and not in the "directions-of-arrival" domain (parameter p domain). For instance, if the directions of the original sources are rotated by $\Delta p = 20°$ so that the new set of directions

$$\underline{p} + \Delta p \underline{1}_5 = [20°, 73.13°, 98.46°, 121.54°, 146.87°]^T$$

is obtained, then this set is not ambiguous.

Furthermore any subset of 4 elements of $\underline{p}$ or $\underline{\check{p}}$ are also ambiguous sets of directions. If the array operates in the presence of 4 sources with directions any subset of 4 elements of $\underline{p}$ will give the same MUSIC spectrum.

It becomes clear that if one ambiguous set is identified, then by simple rotation in arc lengths, an infinite number of ambiguous sets can be generated and therefore two different ambiguous sets may in fact be just a rotation of each other. It is an impossible task to try to identify all the ambiguous sets and, since all these sets can be generated from a single set, the idea of the *ambiguous generator set* arises, defined as follows:

Definition 6.5 — Ambiguous Generator Set: An ordered set $\underline{s} = [s_1, s_2, \ldots, s_c]^T$ of c arc lengths, where $2 \leq c \leq N$, is said to be an *ambiguous generator set (AGS) of arc lengths* if and only if:

(a) $s_1 = 0$ and $s_j \neq 0$ with $2 \leq j \leq c$. That is, all the elements of the set but the first element are non-zero.

(b) the rank of the $N \times c$ matrix $\mathbb{A}(\underline{s})$ with columns the manifold vectors associated with the elements of the set is less than c, i.e.

$$\mathbf{rank}(\mathbb{A}(\underline{s})) \triangleq \rho_a < c \tag{6.20}$$

(c) for any subset $\underline{s}_i$ of m elements of $\underline{s}$ (including $s_1 = 0$) with $\rho_a \leq m < c$, the rank of $\mathbb{A}(\underline{s}_i)$ is equal to ρ_a

According to the previous definition, a set of arc lengths $\underline{s} = [s_1, s_2, s_3, s_4]^T$ with $s_1 = 0$, $s_2, s_3, s_4 \neq 0$ and $\underline{\mathbf{a}}(s_3) = c_1\underline{\mathbf{a}}(s_1) + c_2\underline{\mathbf{a}}(s_2)$ is not an ambiguous generator set. This is because

$$\mathbf{rank}([\underline{\mathbf{a}}(s_1), \underline{\mathbf{a}}(s_2), \underline{\mathbf{a}}(s_3), \underline{\mathbf{a}}(s_4)]) = 3 \tag{6.21}$$

but

$$\mathbf{rank}([\underline{\mathbf{a}}(s_1), \underline{\mathbf{a}}(s_2), \underline{\mathbf{a}}(s_3)]) = 2 \tag{6.22}$$

(i.e. third condition is not satisfied). On the other hand, the subset $[s_1, s_2, s_3]^T$ is an ambiguous generator set since it satisfies all the three conditions of Definition 6.5.

The next section is concerned with the identification of *ambiguous generator sets* (AGS) existing in the manifold of a linear array of arbitrary geometry.

6.3 Partitioning the Array Manifold Curve into Segments of Equal Length

In this section a framework is proposed for the calculation of a type of *ambiguous generator sets of directions* existing in linear arrays of any geometry, as well as their associated *rank of ambiguity*. The framework is based on the uniform partitioning of the array manifold hyperhelix, obtained by dividing the manifold length l_{m} by the difference between any two array sensor locations. Note that from now on the vector $\underline{r}$ with elements the sensor locations (in half-wavelengths) will be assumed to be ordered in the sense that $r_i < r_j \Leftrightarrow i < j$. Furthermore, the absolute difference between the i-th and j-th sensor locations, measured in half-wavelengths, will be denoted by $\Delta r_{ij} > 0$ and hence the aperture of the array will be $\Delta r_{1,N}$. Note that the term "intersensor spacing" will refer to the spacing between any two sensors, not necessarily adjacent.

Thus one type of array ambiguity can be identified by dividing a hyperhelical manifold curve into equal segments, according to the following theorem.

Theorem 6.2 — ***Uniform-Partition Type Ambiguities:*** *Consider a linear array (or an ELA) of N sensors, with locations $\underline{r} = [r_1, r_2, \ldots, r_N]^T$ and let l_{m} denote the length of its hyperhelical manifold curve embedded in*

$\mathfrak{C}^N$. *By uniformly partitioning this hyperhelix into segments of arc length* $\frac{l_m}{\Delta r_{ij}}$ *where* $\Delta r_{ij} = |r_i - r_j|$ *with* $i \neq j$, *a set of c points on the manifold curve can be identified, forming the vector*

$$\underline{s}_{\Delta r_{ij}} = \left[0, \frac{l_m}{\Delta r_{ij}}, 2\frac{l_m}{\Delta r_{ij}}, \ldots, (c-1)\frac{l_m}{\Delta r_{ij}}\right]^T \tag{6.23}$$

where

$$c = \begin{cases} 1 + \mathbf{fix}(\Delta r_{ij}) & \text{if } \Delta r_{ij} \notin \mathcal{N}^+ \\ \Delta r_{ij} & \text{if } \Delta r_{ij} \in \mathcal{N}^+ \end{cases}$$

and fix $(\bullet)$ *denotes the integer part of a number*

(a) If $c \geq N$ *then* any subset *of* $\underline{s}_{\Delta r_{ij}}$ $\forall i, j$ *with N elements is an ambiguous set.*

(b) if $c < N$ *then* $\underline{s}_{\Delta r_{ij}}$ $\forall i, j$ *might be an ambiguous set.*

Proof. The proof of this theorem is as follows. Consider a linear array of N sensors with locations $\underline{r}$. If the difference between the ith and jth sensor locations, i.e. $\Delta r_{ij} = |r_i - r_j|$, satisfies $c \geq N$, then we have to prove that any subset of N elements of the set of arc lengths $\underline{s}_{\Delta r_{ij}}$ of Eq. (6.23) is an ambiguous set.

Using the manifold vector expressed in terms of arc length, i.e.

$$\underline{\mathbf{a}}(s) = \exp(j(\underline{\widetilde{r}}s - \pi\underline{r} + \underline{v})) \tag{6.24}$$

in conjunction with

$$l_m = 2\pi \|\underline{r}\| \quad \text{and} \quad \underline{\widetilde{r}} = \frac{\underline{r}}{\|\underline{r}\|} \tag{6.25}$$

the matrix $\mathbb{A}(\underline{s}_{\Delta r_{i,j}}) \in \mathfrak{C}^{N \times c}$ with columns the associated manifold vectors can be written as follows:

$$\begin{aligned} \mathbb{A}(\underline{s}_{\Delta r_{ij}}) = \Big[&\exp\left(j\left(-\pi\underline{r} + \underline{v}\right)\right), \exp\left(j\left(\tfrac{2\pi}{\Delta r_{ij}}\underline{r} - \pi\underline{r} + \underline{v}\right)\right), \ldots, \\ &\exp\left(j\left((c-1)\tfrac{2\pi}{\Delta r_{ij}}\underline{r} - \pi\underline{r} + \underline{v}\right)\right)\Big] \end{aligned} \tag{6.26}$$

Since $c \geqslant N$ the number of columns of $\mathbb{A}(\underline{s}_{\Delta r_{ij}})$ is equal to, or greater than, N. Hence, by taking a submatrix of $\mathbb{A}_N$ of $\mathbb{A}$ with exactly N elements, e.g. the N first columns of $\mathbb{A}$, we have the following square submatrix denoted by $\mathbb{A}_N$

$$\begin{aligned}\mathbb{A}_N &= \Big[\exp\left(j\left(-\pi\underline{r}+\underline{v}\right)\right), \exp\left(j\left(\tfrac{2\pi}{\Delta r_{ij}}\underline{r}-\pi\underline{r}+\underline{v}\right)\right), \ldots, \\ &\quad \exp\left(j\left((c-1)\tfrac{2\pi}{\Delta r_{ij}}\underline{r}-\pi\underline{r}+\underline{v}\right)\right)\Big] \\ &= \mathbf{diag}(\exp\left(j\left(-\pi\underline{r}+\underline{v}\right)\right)\ \widetilde{\mathbb{A}}_N \end{aligned} \tag{6.27}$$

where

$$\widetilde{\mathbb{A}}_N = \left[\underline{1}_N, \exp\left(j2\pi\frac{1}{\Delta r_{ij}}\underline{r}\right), \ldots, \exp\left(j(2\pi(N-1)\frac{1}{\Delta r_{ij}}\underline{r}\right)\right] \tag{6.28}$$

Consider now the ith and jth rows of $\widetilde{\mathbb{A}}_N$ with $i > j$. By using the property that

$$\exp\left(-j2\pi k\frac{\Delta r_{ij}}{\Delta r_{ij}}\right) = 1, \forall k \in \mathcal{N}^+ \tag{6.29}$$

the ith row of $\widetilde{\mathbb{A}}_N$ can be written as follows

$$\begin{aligned} \text{row}_i^T(\widetilde{\mathbb{A}}_N) &= \mathbb{I}_N \cdot \text{row}_i^T(\widetilde{\mathbb{A}}_N) \\ &= \begin{bmatrix} 1 & 0 & 0 & \ldots & 0 \\ 0 & \exp\left(-j2\pi\frac{\Delta r_{ij}}{\Delta r_{ij}}\right) & 0 & \ldots & 0 \\ 0 & 0 & \exp\left(-j2\pi 2\frac{\Delta r_{ij}}{\Delta r_{ij}}\right) & \ldots & 0 \\ 0 & 0 & 0 & \ldots & 0 \\ 0 & 0 & 0 & \ldots & \exp\left(-j2\pi(N-1)\frac{\Delta r_{ij}}{\Delta r_{ij}}\right) \end{bmatrix} \cdot \\ &\quad \left[\underline{1}_N, \exp\left(j2\pi r_i\tfrac{1}{\Delta r_{ij}}\right), \ldots, \exp\left(j2\pi r_i\left(N-1\right)\tfrac{1}{\Delta r_{ij}}\right)\right]^T \\ &= \left[\underline{1}_N, \exp\left(j2\pi\tfrac{r_i-\Delta r_{ij}}{\Delta r_{ij}}\right), \ldots, \exp\left(j2\pi(N-1)\tfrac{r_i-\Delta r_{ij}}{\Delta r_{ij}}\right)\right]^T \\ &= \left[\underline{1}_N, \exp\left(j2\pi\tfrac{r_j}{\Delta r_{ij}}\right), \ldots, \exp\left(j2\pi(N-1)\tfrac{r_j}{\Delta r_{ij}}\right)\right]^T \\ &= \text{row}_j^T(\widetilde{\mathbb{A}}_N) \end{aligned}$$

i.e.

$$\begin{aligned} \text{row}_i(\widetilde{\mathbb{A}}_N) &= \text{row}_j(\widetilde{\mathbb{A}}_N) \\ \Longrightarrow \mathbf{det}\left(\widetilde{\mathbb{A}}_N\right) &= 0 \Longrightarrow \mathbf{det}\left(\mathbb{A}_N\right) = 0 \end{aligned} \tag{6.30}$$

It is easy to see that the fact that $\mathbb{A}_N$ was chosen to consist of the first N columns of $\mathbb{A}$, and not any N columns, is not restrictive in the least since the ith and the jth rows of any $N \times N$ submatrix of $\mathbb{A}$ will be equal. Since all submatrices $\mathbb{A}_N$ of $\mathbb{A}$ are singular, it results that $\mathbb{A}$ is rank deficient and therefore, any subset of $\underline{s}_{\Delta r_{i,j}}$ with exactly N elements is an ambiguous set.

□

Note that

$$c = \begin{cases} 1 + \mathbf{fix}(\Delta r_{ij}) & \text{if } \Delta r_{ij} \notin \mathcal{N}^+ \\ \Delta r_{ij} & \text{if } \Delta r_{ij} \in \mathcal{N}^+ \end{cases} \Longrightarrow c < 1 + \Delta r_{ij} \Longrightarrow (c-1)\frac{l_{\mathrm{m}}}{\Delta r_{ij}} < l_{\mathrm{m}} \tag{6.31}$$

indicating that the last element of the vector $\underline{s}_{\Delta r_{ij}}$ of Eq. (6.23), is smaller that the manifold length l_{m}. The vector $\underline{s}_{\Delta r_{ij}}$ with elements $c \in \mathcal{N}^+$ manifold points will be known as the *"uniform basic set" (UBS) of arc lengths associated with the intersensor spacing* Δr_{ij} as it is constructed by uniformly partitioning the manifold length by the spacing Δr_{ij} between the ith and jth sensors.

According to Theorem 6.2, all subsets of N elements from a set $\underline{s}_{\Delta r_{ij}}$, of the form of Eq. (6.23) are ambiguous sets but not necessarily *ambiguous generator sets.* In order for such a subset to be an ambiguous generator set it must have its first element equal to zero and also satisfy the third condition of the ambiguous generator set in Definition 6.5.

From the investigation of the precise cause of the ambiguous nature of the set of arc lengths $\underline{s}_{\Delta r_{ij}}$ constructed by uniformly partitioning the manifold length by the spacing Δr_{ij}, it is clear that the most common AGSs are of rank $N-1$. However, AGSs of rank less than $(N-1)$ can exist if more pairs of rows of the determinant matrix $\widetilde{\mathbb{A}}_N(\underline{s}_{\Delta r_{ij}})$ are identical. This is possible if the array geometrical structure satisfies certain special conditions and for this reason we need the following definition.

Definition 6.6 — A "Distinct" Set of Intersensor Spacings: In a linear array of N sensors, a set of k intersensor spacings is said to be a "distinct" set if no intersensor spacing is a combination of the other spacings in the set. There are $N(N-1)/2$ intersensor spacings but at most $(N-1)$ elements in a distinct set of intersensor spacings (i.e. $k \leqslant N-1$)).

For instance, if Δr_{ij} and Δr_{jk} are members of a distinct set with $i \neq j \neq k$ then Δr_{ik} is not a member.

Based on the above definition, the following theorem states that if there are distinct sets of intersensor spacings then AGSs of rank less than $(N-1)$ may be found.

Theorem 6.3 — Number of AGSs of a given rank: *Consider an N-sensor linear array. If $k_{\Delta r_{ij}}$ denotes the elements of a distrinct set of intersensor spacings which are integer multiples of the smallest (Δr_{ij}) amongst them, and*

$$\Delta r_{ij} \geqslant \left(N - k_{\Delta r_{ij}}\right) \tag{6.32}$$

then the Uniform Basic Set $\underline{s}_{\Delta r_{ij}}$ of c elements is an ambiguous set from which

$$\sum_{\ell=N-k_{\Delta r_{ij}}}^{\min(N-1,c-1)} \binom{\mathbf{c}-\mathbf{1}}{\ell} \tag{6.33}$$

ambiguous generator sets of rank-$\left(N - k_{\Delta r_{ij}}\right)$ can be constructed.

The implication of the above theorem to the number of ambiguous generator sets of a linear array is twofold. Firstly, it reveals the existence of additional ambiguous generator sets corresponding to intersensor spacings which are less than $(N-1)$, provided that they occur in integer multiples of each other, which includes the special case of repeated occurrences. In particular, if a spacing, say Δr_{ij}, occurs $k_{\Delta r_{ij}}$ times then the size of this spacing need only be $\left(N - k_{\Delta r_{ij}}\right)$ to give rise to ambiguous generator sets. Secondly, the larger the $k_{\Delta r_{ij}}$ is, the more rank-$\left(N - k_{\Delta r_{ij}}\right)$ ambiguous generator sets can be constructed. This point can be illustrated from the intersensor spacings $\Delta r_{13} = \Delta r_{24} = 4.1$ of the linear array $\underline{r} = [-3, -2.2, 1.1, 1.9, 2.2]^T$ where $k_{\Delta r_{13}} = 2$ and $\Delta r_{13} \geqslant (N - k_{\Delta r_{13}}) = 3$. Thus from $\underline{s}_{\Delta r_{1,3}}$, five rank-3 ambiguous generator sets can be constructed (instead of a single rank-4 set).

From the array's geometrical perspective, the following statement with respect to rank-$\left(N - k_{\Delta_{r_{ij}}}\right)$ ambiguity can be made:

Corollary 6.1 *If k intersensor spacings of an N-sensor linear array are integer multiples of the smallest among them, with the latter being at least $(N-k)$ half-wavelengths, then the array suffers from rank-$(N-k)$ ambiguity.*

In summary the reason for the existence of these lower than $N-1$ rank ambiguities can easily be deduced from the definition of the ambiguous generator set and especially its third condition (see Definition 6.5). More analytically, it can be easily derived from the proof of Theorem 2.1, that if two or more intersensor spacings of a linear array (or ELA) are integer multiples of each other (which is the case for either of the above conditions) and if $\underline{s}$ is an ambiguous generator set associated with the smallest of these intersensor spacings, then the manifold matrix $\mathbb{A}(\underline{s})$ may have more than two identical rows. This implies that its rank will be smaller than $N-1$.

As a direct result of Theorem 6.3 and of the previous corollary it can be said that the uniform linear array (or ELA) with intersensor spacing $\Delta r_{1,2} \geqslant 1$ suffers from trivial ambiguities. Indeed, the intersensor spacing $\Delta r = \Delta r_{1,2}$ occurs $N-1$ times. That is $\Delta r_{1,2} = \Delta r_{2,3} = \cdots = \Delta r_{N-1,N}$, i.e.

$$k_{\Delta_{r_{ij}}} = N-1 \Longrightarrow \rho_{\mathrm{a}} = N-N+1 = 1 \tag{6.34}$$

This well known result is rediscovered by observing that in such a case the set $\underline{s}_{\Delta r_{i,j}}$ will contain at least two elements and the matrix $\mathbb{A}$ will be rank deficient since all its rows will be equal.

By setting the value of $\Delta r_{ij} = |r_i - r_j|$ in the above equation to be equal to the biggest possible difference between two array sensor locations, which is obviously the aperture $l_{\mathrm{a}} = \Delta r_{1,N}$, and by combining the above with the Theorem 6.2(a), the following sufficient condition can be obtained.

Corollary 6.2 *A SUFFICIENT condition for the presence of ambiguities in ALL linear arrays of N sensors, is*

$$l_{\mathrm{a}} > N-1 \quad \textit{where} \quad l_{\mathrm{a}} = \max_{\forall i,j} \Delta r_{ij} = \Delta r_{1,N} \tag{6.35}$$

It should be clear however, that the above provides a sufficient but by no means necessary condition for the presence of ambiguities. This means that a linear array (or ELA) can possibly suffer from ambiguities even if $l_{\mathrm{a}} \leq N-1$ half wavelengths. Only for the specific case of uniform linear (or ELA) arrays, has it been proven that no ambiguities exist if $l_{\mathrm{a}} \leq N-1$ or, equivalently, if the intersensor spacing is not greater than half a wavelength.

Theorem 6.2 has an implication that needs to be stressed. It is well known that an increase in the aperture results in better resolution capabilities. However, from the point of view of ambiguities, increasing the aperture may not be a very good idea. By increasing the aperture, the number of elements in the set of Eq. (6.23) increases, and once this number becomes greater than $N-1$, the set becomes ambiguous. Furthermore, increasing all the intersensor spacings may be considered an even more serious issue since an array with sensor locations given by $\check{\underline{r}} = b\underline{r}$, with $b > 1$, might have many intersensor spacings Δr_{ij} which result in ambiguous sets of the form given in Eq. (6.23).

Example 6.2 Consider an array with sensor locations given by $\underline{r} = [-2.1, -0.9, 0.9, 2.1]^T$. The manifold length l_{m} of this array is calculated to be 20.3016. Furthermore, the only absolute difference between sensor locations satisfying $c \geqslant N$ is the aperture $\Delta r_{1,4} = 4.2$. This provides the following uniform basic set of arc lengths:

$$\underline{s}_{\Delta r_{1,4}} = [0, 4.8337, 9.6674, 14.5011, 19.3348]^T$$

According to Theorem 6.2, any subset of 4 elements ($N = 4$) from the above set will be an ambiguous set and the matrix $\mathbb{A}$ with columns the manifold vectors corresponding to such a subset will be singular. Furthermore, it can be shown that no matrix $\mathbb{A}$ corresponding to any three elements of $\underline{s}_{\Delta r_{1,4}}$ is rank deficient. This means that four different ambiguous generator sets can be identified from $\underline{s}_{\Delta r_{1,4}}$ which are all the subsets of $\underline{s}_{\Delta r_{1,4}}$ with their first element zero and three non-zero elements of $\underline{s}_{\Delta r_{1,4}}$. As expected from Theorem 6.2, the set $\underline{s}_{\Delta r_{1,4}}$ is ambiguous since $c \geq N$. However, ambiguous sets might also be defined from Theorem 6.2 when $c < N$. To see this, consider the uniform basic set for the difference $\Delta r_{1,3} = \Delta r_{2,4} = 3$ (i.e. $k_{\Delta r_{1,3}} = 2$), which is $c = 3 < N = 4$

$$\begin{aligned} \underline{s}_{\Delta r_{1,3}} &= \underline{s}_{\Delta r_{2,4}} = [0, 6.7672, 13.5344]^T \\ \Longrightarrow \underline{\theta}_{\Delta r_{1,3}} &= \underline{\theta}_{\Delta r_{2,4}} = [0°, 70.53°, 109.47°]^T \end{aligned}$$

If the matrix $\mathbb{A}$ with columns the corresponding manifold vectors is obtained, then $\mathbf{rank}(\mathbb{A}) = 2$ and hence the above set is an ambiguous set with rank of ambiguity equal to 2. It should also be noted that not all the uniform basic sets $\underline{s}_{\Delta r_{i,j}}$ resulting from a difference Δr_{ij} between two sensor locations are ambiguous if $c < N$. To see this, consider the differences $\Delta r_{1,2} = \Delta r_{3,4} = 1.2$ (or $\Delta r_{2,3} = 1.8$) of the previous array which results in the set $\underline{s}_{\Delta r_{1,2}} = [0, 16.9180]^T$ (or $\underline{s}_{\Delta r_{2,3}} = [0, 11.2787]^T$) (i.e. $c = 2 < N = 4$) which is unambiguous.

So far it has been shown that Theorem 6.2 can be used in order to identify ambiguous sets inherent in linear arrays of any geometry. When this theorem is focused on specific array geometries it can produce some more useful results as the following corollaries indicate.

Corollary 6.3 *All the ambiguous sets that exist in a two-element array can be calculated from Eq. (6.23).*

Proof. The proof is as follows. Consider the two-element array $r = [-r_1, r_1]$. Let $\underline{s} = [s_1, s_2]^T$ with $s_1 = 0$ being an ambiguous set of arc lengths. The matrix $\mathbb{A}$ with columns the manifold vectors corresponding to $\underline{s}$ is

$$\begin{aligned}\mathbb{A} &= [\underline{\mathbf{a}}(0), \underline{\mathbf{a}}(s_2)] \\ &= \begin{bmatrix} \exp(j\pi r_1) & \exp\left(j\pi r_1\left(1 - \dfrac{s_2}{\pi\,\|\underline{r}\|}\right)\right) \\ \exp(-j\pi r_1) & \exp\left(-j\pi r_1\left(1 - \dfrac{s_2}{\pi\,\|\underline{r}\|}\right)\right)\end{bmatrix}\end{aligned} \tag{6.36}$$

Since it is assumed that $\underline{s}$ is an ambiguous set, it results that $\mathbb{A}$ is singular and therefore

$$\begin{aligned}&\det(\mathbb{A}) = \exp(j\pi r_1).\exp\left(-j\pi r_1\left(1 - \frac{s_2}{\pi\|\underline{r}\|}\right)\right) \\ &\quad - \exp(-j\pi r_1).\exp\left(j\pi r_1\left(1 - \frac{s_2}{\pi\|\underline{r}\|}\right)\right) = 0 \\ &\Longrightarrow \exp\left(j\pi r_1 \frac{s_2}{\pi\|\underline{r}\|}\right) - \exp\left(-j\pi r_1 \frac{s_2}{\pi\|\underline{r}\|}\right) = 0 \\ &\Longrightarrow \sin\left(r_1 \frac{s_2}{\|\underline{r}\|}\right) = 0 \\ &\Longrightarrow r_1 \frac{s_2}{\|\underline{r}\|} = k\pi \quad \text{with} \quad k = 1, 2, 3, \ldots \text{ (since } s_2 r_1 \neq 0) \\ &\Longrightarrow s_2 = k\frac{\pi\|\underline{r}\|}{r_1} = k\frac{2\pi\|\underline{r}\|}{2r_1} = k\frac{2\pi\|\underline{r}\|}{\Delta r_{1,2}} = k\frac{l_{\mathrm{m}}}{\Delta r_{1,2}}\end{aligned}$$

That is

$$s_1 = 0 \quad \text{and} \quad s_2 = k\frac{l_{\mathrm{m}}}{\Delta r_{1,2}} \tag{6.37}$$

with

$$k = 1, 2, 3, \ldots$$

Therefore the only ambiguous set of arc lengths that can possibly exist for a two-element array are of the form of Eq. (6.23). As a result, a necessary and sufficient condition for the presence of ambiguities in a two element linear array is

$$\frac{l_{\mathrm{m}}}{\Delta r_{12}} < l_{\mathrm{m}} \Longrightarrow \Delta r_{12} > 1 \tag{6.38}$$

□

Corollary 6.4 *The set of arc lengths* $\underline{s} = [0, \pi\sqrt{2}, 2\pi\sqrt{2}, \ldots, (c-1)\pi\sqrt{2}]^T$ *is an ambiguous Uniform Basic Set for all three-element symmetric linear arrays, as long as* $\Delta r_{1,3} > 2$.

Proof. For all three-element symmetric linear arrays with sensor positions given by $\underline{r} = [-r_1, 0, r_1]^T$, the manifold length is equal to

$$l_{\mathrm{m}} = 2\pi\|\underline{r}\| = 2\pi\sqrt{(-r_1)^2 + (r_1)^2} = 2\pi r_1\sqrt{2} \tag{6.39}$$

The Uniform Basic Set $\underline{s}_{\Delta r_{1,3}}$ associated with the difference between the first and the third sensors is

$$\begin{aligned}\underline{s} &= \left[0, \tfrac{l_{\mathrm{m}}}{\Delta r_{1,3}}, 2\tfrac{l_{\mathrm{m}}}{\Delta r_{1,3}}, \ldots, (c-1)\tfrac{l_{\mathrm{m}}}{\Delta r_{1,3}}\right]^T \\ &= \left[0, \tfrac{2\pi r_1\sqrt{2}}{2r_1}, 2\tfrac{2\pi r_1\sqrt{2}}{2r_1}, \ldots, (c-1)\tfrac{2\pi r_1\sqrt{2}}{2r_1}\right]^T \\ &= \left[0, \pi\sqrt{2}, 2\pi\sqrt{2}, \ldots, (c-1)\pi\sqrt{2}\right]^T \end{aligned}\tag{6.40}$$

□

Corollary 6.5 *Let two arrays of N sensors have a common difference* Δr *between two sensor locations. The Uniform Basic Set of arc lengths associated with the common* Δr *is different for these two arrays but the corresponding ambiguous sets of DOA's are the same.*

Proof. Consider two different linear arrays (or ELAs) each having N sensors with sensor locations $\underline{r}_1$ and $\underline{r}_2$. Let there be a common difference Δr between two sensor locations, which, for both arrays, satisfy the conditions of Theorem 6.2. In this case, the two Uniform Basic Sets will be:

First array:

$$\underline{s}_1 = \left[0, \tfrac{2\pi\|\underline{r}_1\|}{\Delta r}, 2\tfrac{2\pi\|\underline{r}_1\|}{\Delta r}, \ldots, (c-1)\tfrac{2\pi\|\underline{r}_1\|}{\Delta r}\right]^T \tag{6.41}$$

Second array:

$$\underline{s}_2 = \left[0, \tfrac{2\pi\|\underline{r}_2\|}{\Delta r}, 2\tfrac{2\pi\|\underline{r}_2\|}{\Delta r}, \ldots, (c-1)\tfrac{2\pi\|\underline{r}_2\|}{\Delta r}\right]^T \tag{6.42}$$

with $c \geqslant N$, which implies that the above two sets are ambiguous. Using Eq. (2.28) the corresponding set of DOA's for the first array is

$$\begin{aligned}\underline{\theta}_1 &= \left[0, \arccos\left(1 - \frac{\frac{2\pi\|\underline{r}_1\|}{\Delta r}}{\pi\|\underline{r}_1\|}\right), \ldots, (c-1)\arccos\left(1 - \frac{\frac{2\pi\|\underline{r}_1\|}{\Delta r}}{\pi\|\underline{r}_1\|}\right)\right]^T \\ &= \left[0, \arccos\left(1 - \tfrac{2}{\Delta r}\right), \ldots, (c-1)\arccos\left(1 - \tfrac{2}{\Delta r}\right)\right]^T \end{aligned}\tag{6.43}$$

whereas, for the second array

$$\underline{\theta}_2 = \left[0, \arccos\left(1 - \frac{\frac{2\pi\|\underline{r}_2\|}{\Delta r}}{\pi\|\underline{r}_2\|}\right), \ldots, (c-1)\arccos\left(1 - \frac{\frac{2\pi\|\underline{r}_2\|}{\Delta r}}{\pi\|\underline{r}_2\|}\right)\right]^T$$

$$= \left[0, \arccos\left(1 - \tfrac{2}{\Delta r}\right), \ldots, (c-1)\arccos\left(1 - \tfrac{2}{\Delta r}\right)\right]^T \tag{6.44}$$

The two sets are obviously identical. □

6.3.1 *Calculation of Ambiguous Generator Sets of Linear (or ELA) Array Geometries*

For a given linear array (or ELA) with sensor locations given by $\underline{r}$ and based on Theorem 6.2, the following technique (presented in a step by step form) provides a table with rows all the ambiguous generator sets, and their associated rank of ambiguity:

(1) For a given array with sensor locations $\underline{r}$, calculate the manifold length $l_{\mathrm{m}} = 2\pi\|\underline{r}\|$.

(2) Calculate the intersensor spacings by taking the Kronecker difference between the vector $\underline{r}$ (with elements the sensor locations) and itself, i.e. $\underline{r} \ominus \underline{r}$, which results in a $(N^2 \times 1)$-dimensional real vector. Create a new vector $\underline{\Delta r}$ by keeping all the positive elements of $\underline{r} \ominus \underline{r}$. Due to the properties of the Kronecker difference N elements of $\underline{r} \ominus \underline{r}$ will be equal to zero and half of the remaining elements will be negative. Therefore, $\underline{\Delta r} \in \mathcal{R}^{d\times 1}$, where $d = \frac{N}{2}(N-1)$.

(3) Redefine $\underline{\Delta r}$ by eliminating all elements which are smaller than one, i.e. $d \leq \frac{N}{2}(N-1)$. Note that the elements of $\underline{\Delta r}$ which are smaller than one provide

$$c = 1 + \mathrm{fix}(\Delta r_{ij}) = 1 + 0 = 1 \tag{6.45}$$

and therefore cannot possibly give rise to ambiguous generator sets.

(4) For each element of $\underline{\Delta r}$, i.e. for each difference $\Delta r_{ij} = |r_i - r_j| > 1$ between two sensor locations, construct the "Uniform Basic Set" vector $\underline{s}_{\Delta r_{ij}}$ using Eq. (6.23). Note that there are d different vectors $\underline{s}_{\Delta r_{ij}}$ and that each $\underline{s}_{\Delta r_{i,j}}$ corresponds to a different uniform partition of the array manifold.

(5) Identify all the vectors $\underline{s}_{\Delta r_{ij}}$, with $c < N$ and then eliminate those for which:

$$\mathrm{rank}(\mathbb{A}(\underline{s}_{\Delta r_{ij}})) = c < N$$

Form the "Table of Ambiguous Uniform Basic Sets" with rows all the remaining vectors $\underline{s}_{\Delta r_{ij}}$. All these vectors are ambiguous sets and produce ambiguous generator sets. It is obvious that the set with the maximum number of elements is the set $\underline{s}_{\Delta r_{1,N}}$.

(6) For each $\underline{s}_{\Delta r_{ij}}$ (i.e. for each row of the "Table of Uniform Basic Sets") construct the AGSs (and associated rank of ambiguity), according to the following rules:

Rule (a) If the non-zero elements of $\underline{s}_{\Delta r_{ij}}$ cannot be found in any other "Uniform Basic Set" vector, then

$$\binom{c-1}{N-1} \text{ ambiguous generator sets}$$

can be produced by this $\underline{s}_{\Delta r_{ij}}$. These ambiguous generator sets are all the possible subsets of N elements of $\underline{s}_{\Delta r_{i,j}}$ with their first element zero and $N-1$ non-zero elements of $\underline{s}_{\Delta r_{ij}}$.

> **All ambiguous generator sets constructed in this way have rank of ambiguity**
> $$\rho_{\mathrm{a}} = N - 1$$

Note that such rows must definitely have $c \geq N$ (see Theorem 6.2), otherwise they would have been eliminated in Step 5.

Rule (b) If the non-zero elements of $\underline{s}_{\Delta r_{i,j}}$ can be found in other rows, then **ambiguous generator sets with rank of ambiguity less than** $N-1$ **might be obtained**. This means that all subsets of $\underline{s}_{\Delta r_{i,j}}$ with their first element 0 and with lengths $2, 3, \ldots, \mathbf{min}(N-1, c-1)$ must be considered. These subsets are classified as ambiguous generator sets if the three conditions of the ambiguous generator set definition (see Definition 6.5) are satisfied. Furthermore, for each ambiguous generator set $\underline{s}$, rank of ambiguity $\rho_{\mathrm{a}} = \mathrm{rank}(\mathbb{A}(\underline{s}_{\Delta r_{i,j}}))$ is estimated. Note that this step clarifies why in Step 5 it is incorrect to eliminate two $\underline{s}_{\Delta r_{i,j}}$ vectors which are the same, although they will result in the same ambiguous generator sets.

(7) Create the Ambiguous Generator Table of the array whose rows are all the different ambiguous generator sets (and their associated rank

of ambiguity) found in Step 6. Also, eliminate rows of $\underline{s}_{\Delta r_{i,j}}$ which are duplicates.

6.4 Representative Examples

An indicative example of the above technique is presented below.

Example 6.3 The steps of the previously described technique for an array of $N = 5$ sensors with sensor positions $\underline{r} = [-3, -2.2, 1.1, 1.9, 2.2]^T$ are as follows:

(1) The manifold length of this array is computed to be $l_m = 30.4589$.
(2) The Kronecker difference between $\underline{r}$ and itself, reshaped in the form of an $N \times N$ matrix, is

$$\begin{bmatrix} 0, & -0.8, & -4.1, & -4.9, & -5.2, \\ 0.8, & 0, & -3.3 & -4.1 & -4.4, \\ 4.1, & 3.3, & 0, & -0.8, & -1.1 \\ 4.9, & 4.1, & 0.8, & 0, & -0.3 \\ 5.2, & 4.4, & 1.1, & 0.3, & 0 \end{bmatrix}$$

(3) Eliminating those entries of $\underline{r} \ominus \underline{r}$ which are smaller than unity results in

$$\underline{\Delta r} = \begin{bmatrix} \Delta r_{1,3} = 4.1 \\ \Delta r_{1,4} = 4.9 \\ \Delta r_{1,5} = 5.2 \\ \Delta r_{2,3} = 3.3 \\ \Delta r_{2,4} = 4.1 \\ \Delta r_{2,5} = 4.4 \\ \Delta r_{3,5} = 1.1 \end{bmatrix} \tag{6.46}$$

(4) The seven Uniform Basic Sets (vectors) $\underline{s}_{\Delta r_{i,j}}$ are given in Table 6.2.
(5) The Uniform Basic Sets $\underline{s}_{\Delta r_{ij}}$ which have less than $N = 5$ elements (i.e. $c < N$) are examined. These are the sets $\underline{s}_{\Delta r_{2,3}}$ and $\underline{s}_{\Delta r_{3,5}}$.
    - The number of elements in $\underline{s}_{\Delta r_{2,3}}$ is $4 < N = 5$ and, furthermore, rank $(\mathbb{A}(\underline{s}_{\Delta r_{2,3}})) = 4 = c$ and hence this row is eliminated.
    - Also, $\underline{s}_{\Delta r_{3,5}}$ is eliminated since $c = 2 < N = 5$ and

$$\mathbf{rank}(\mathbb{A}(\underline{s}_{\Delta r_{2,3}})) = 2 = c. \tag{6.47}$$

Thus the remaining sets are $\underline{s}_{\Delta r_{1,3}}, \underline{s}_{\Delta r_{1,4}}, \underline{s}_{\Delta r_{1,5}}, \underline{s}_{\Delta r_{2,4}}$ and $\underline{s}_{\Delta r_{2,5}}$.

(6) In this step, the AGSs arising from the remaining Uniform Basic Sets are calculated based on the rules (6a) and (6b).

Table 6.2 Uniform Basic Sets of $\Delta r_{ij} > 1$.

	s_1	s_2	s_3	s_4	s_5	s_6
$\underline{s}^T_{\Delta r_{1,3}}$	0	7.4290	14.8580	22.2870	29.7160	—
$\underline{s}^T_{\Delta r_{1,4}}$	0	6.2161	12.4322	18.6483	24.8644	—
$\underline{s}^T_{\Delta r_{1,5}}$	0	5.8575	11.7150	17.5725	23.4300	29.2875
$\underline{s}^T_{\Delta r_{2,3}}$	0	9.2300	18.4600	27.6900	—	—
$\underline{s}^T_{\Delta r_{2,4}}$	0	7.4290	14.8580	22.2870	29.7160	—
$\underline{s}^T_{\Delta r_{2,5}}$	0	6.9225	13.8450	20.7675	27.6900	—
$\underline{s}^T_{\Delta r_{3,5}}$	0	27.6900	—	—	—	—

The non-zero elements of $\underline{s}_{\Delta r_{1,4}}, \underline{s}_{\Delta r_{1,5}}$ and $\underline{s}_{\Delta r_{2,5}}$ cannot be found in any other $\underline{s}_{\Delta r_{i,j}}$ and hence should be investigated based on rule 6a. The non-zero elements of $\underline{s}_{\Delta r_{1,3}}$ can also be found in $\underline{s}_{\Delta r_{2,4}}$ and vice-versa and hence should be investigated based on rule 6b. The above are summarized in Table 6.3.

Table 6.3 Rules for examining the UBSs of Table 6.2.

	c	ρ_{a}	rule
$\underline{s}^T_{\Delta r_{1,3}}$	5	3	(6b)
$\underline{s}^T_{\Delta r_{1,4}}$	5	4	(6a)
$\underline{s}^T_{\Delta r_{1,5}}$	6	4	(6a)
$\underline{s}^T_{\Delta r_{2,3}}$	4	4	*removed*
$\underline{s}^T_{\Delta r_{2,4}}$	5	3	(6b)
$\underline{s}^T_{\Delta r_{2,5}}$	5	4	(6a)
$\underline{s}^T_{\Delta r_{3,5}}$	2	2	*removed*

Based on Rule (6a)

- $\underline{s}_{\Delta r_{1,4}}$: Since the number of the elements of $\underline{s}_{\Delta r_{1,4}}$ is 5 ($c = 5$), we have $\binom{c-1}{N-1} = \binom{4}{4} = 1$ which implies that only one ambiguous generator set (which is the same as the Uniform Basic Set $\underline{s}_{\Delta r_{1,4}}$) can be identified, with $\rho_{\mathrm{a}} = 4$.
- $\underline{s}_{\Delta r_{1,5}}$: This time $c = 6$ and therefore $\binom{c-1}{N-1} = \binom{5}{4} = 5$ ambiguous generator sets can be found by taking all the subsets of N elements of $\underline{s}_{\Delta r_{1,5}}$ (with one zero and $N-1$ non-zero elements). Each of these sets has $\rho_{\mathrm{a}} = 4$.

- $\underline{s}_{\Delta r_{2,5}}$: Finally, $\underline{s}_{\Delta r_{2,5}}$ has $c = 5$ and it generates $\binom{c-1}{N-1} = \binom{4}{4} = 1$ ambiguous generator set. This AGS is the same as set $\underline{s}_{\Delta r_{2,5}}$ and has $\rho_a = 4$.

Based on Rule (6b)

- $\underline{s}_{\Delta r_{1,3}}$: The non-zero elements of $\underline{s}_{\Delta r_{1,3}}$ can also be found in $\underline{s}_{\Delta r_{2,4}}$ and hence $\underline{s}_{\Delta r_{1,3}}$ is investigated based on rule 6b. The ambiguous generator set definition is initially checked for the set $\underline{s}_{\Delta r_{1,3}}$. This set is certainly ambiguous and the matrix with columns the corresponding manifold vectors has

$$\text{rank}\left(\mathbb{A}(\underline{s}_{\Delta r_{1,3}})\right) = 3 < c = 5 \tag{6.48}$$

However, any one of the four subsets $\check{\underline{s}}$ with four elements of $\underline{s}_{\Delta r_{1,3}}$ involving one zero and three non-zero elements, is also an ambiguous set. The matrix with columns the corresponding manifold vectors has $\text{rank}(\mathbb{A}(\check{\underline{s}})) = 3 < c = 5$. Therefore, the set $\underline{s}_{\Delta r_{1,3}}$ satisfies the third condition of the definition of the ambiguous generator set. Similarly, the four subsets $\check{\underline{s}}$ are ambiguous generator sets since all their subsets with three elements are unambiguous. Hence, the following five ambiguous generator sets can be defined from $\underline{s}_{\Delta r_{1,3}}$, all of which have $\rho_a = 3$. Indeed

$$\begin{aligned}\sum_{\ell=N-k_{\Delta r_{ij}}}^{\min(N-1,c-1)} \binom{c-1}{\ell} &= \sum_{\ell=5-2}^{\min(4,4)} \binom{5-1}{\ell} \\ &= \sum_{\ell=3}^{4} \binom{4}{\ell} = \binom{4}{3} + \binom{4}{4} = 5 \end{aligned} \tag{6.49}$$

- $\underline{s}_{\Delta r_{2,4}}$: Since $\underline{s}_{\Delta r_{2,4}}$ is the same as $\underline{s}_{\Delta r_{1,3}}$, it follows that the ambiguous generator sets obtained from this row are the ones which were already obtained.

(7) The ambiguous generator table, Table 6.4, with rows the AGSs produced in Step-6 is now constructed. Their associated rank of ambiguity is also shown.

In conclusion, twelve ambiguous generator sets can be identified for this array. Five of these have $\rho_a = 3$, while the remaining have $\rho_a = 4$, constructed from UBSs as shown below:

AGS	UBS
1st to 5th	$\underline{s}_{\Delta r_{1,3}}; \underline{s}_{\Delta r_{2,4}}$
6th	$\underline{s}_{\Delta r_{1,4}}$
7th to 11th	$\underline{s}_{\Delta r_{1,5}}$
12th	$\underline{s}_{\Delta r_{2,5}}$

Table 6.4 List of AGSs and their associated rank of ambiguity.

	s_1	s_2	s_3	s_4	s_5	ρ_a
(1)	0	7.4290	14.8580	22.2870	29.7160	3
(2)	0	7.4290	14.8580	22.2870	—	3
(3)	0	7.4290	14.8580	29.7160	—	3
(4)	0	7.4290	22.2870	29.7160	—	3
(5)	0	14.8580	22.2870	29.7160	—	3
(6)	0	6.2161	12.4322	18.6483	24.8644	4
(7)	0	5.8575	11.7150	17.5725	23.4300	4
(8)	0	5.8575	11.7150	17.5725	29.2875	4
(9)	0	5.8575	11.7150	23.4300	29.2875	4
(10)	0	5.8575	17.5725	23.4300	29.2875	4
(11)	0	11.7150	17.5725	23.4300	29.2875	4
(12)	0	6.9225	13.8450	20.7675	27.6900	4

6.5 Handling Ambiguities in Planar Arrays

The material presented in this chapter is valid for any hyperhelical curve (of the form of Eq. (6.2)) embedded in $\mathfrak{C}^N$. However, for *p*-curves $\mathcal{A}_{p|q_o}$ ($p = \phi, \alpha$ or β) some extra steps should be taken to accommodate the property that these curves of hyperhelical shape are lying on the manifold surface of a planar array. These curves can be seen as the array manifolds of an ELA with location given by $\underline{r}(\Psi_0)$ represented by Eqs. (6.3) and (6.4). Therefore, the Uniform Basic Sets $\underline{s}_{\Delta r_{ij}}$ (see Eq. (6.23)), and consequently the AGSs, can be constructed according to the procedure described in the previous section with

$$\begin{aligned}\underline{s}_{\Delta r_{ij}} &\triangleq \underline{s}_{\Delta r_{ij}}(\Psi_0)\\ &= \left[0,\ \frac{2\pi\|\underline{r}(\Psi_0)\|}{\Delta r_{ij}},\ 2\frac{2\pi\|\underline{r}(\Psi_0)\|}{\Delta r_{ij}}, \ldots, (c-1)\frac{2\pi\|\underline{r}(\Psi_0)\|}{\Delta r_{ij}}\right]^T \qquad (6.50)\\ &\forall i,j \quad \text{with} \quad i \neq j\end{aligned}$$

where Δr_{ij} represents the intersensor spacing between the ith and jth sensors of the ELA $\underline{r}(\Psi_0)$.

Before we proceed it is important to point out that

(1) ambiguous sets of directions with a common azimuth are located on ϕ-curves, while
(2) ambiguous sets of directions of different azimuth and elevation angles are located on the α-curves (or β-curves) for a rotation Θ.

6.5.1 *Ambiguities on ϕ-curves*

Equation (4.24) of Chapter 4 which is reshown below

$$\underline{\mathbf{a}}(\phi \mid \theta_o) = \underline{\mathbf{a}}(\phi \mid \theta_o + 180°) \tag{6.51}$$

implies that the two ϕ-curves $\mathcal{A}_{\phi|\theta_o}$ and $\mathcal{A}_{\phi|\theta_o+180°}$ have the same length and curvatures and can be considered as a continuation of one another forming a composite ϕ-curve having a hyperhelical shape. The family of these composite curves (from now on known as φ-curves) can be described with an alternative but equivalent parameter space as

$$\Omega = \{(\vartheta_o, \varphi) : \vartheta_o \in [0°, 180°), \quad \varphi \in [0°, 180°)\} \tag{6.52}$$

i.e.

$$\Omega_\vartheta = \Omega_\varphi = [0°, 180°) \tag{6.53}$$

and is defined as

$$\{\mathcal{A}_{\varphi|\vartheta_o}, \ \forall \vartheta_o : \vartheta_o \in \Omega_\vartheta\} \tag{6.54}$$

where

$$\mathcal{A}_{\varphi|\vartheta_o} = \{\underline{\mathbf{a}}(\vartheta_o, \varphi) \in \mathfrak{C}^N, \ \forall \varphi : \varphi \in \Omega_\varphi, \quad \vartheta_o = \text{constant}\} \tag{6.55}$$

It is obvious now that a φ-curve represents directly the array manifold of an Equivalent Linear Array (ELA) with sensor locations given by

$$\underline{r}(\Psi_0), \quad \text{with} \quad \Psi_0 \triangleq \vartheta_o \tag{6.56}$$

having a manifold length

$$(l_m \text{ of } \mathcal{A}_{\varphi|\vartheta_o}) = 2\pi \|\underline{r}(\vartheta_o)\| \tag{6.57}$$

This implies that every member of the family of φ-curves of a planar array has its own ELA. Based on the above discussion the transformation of the parameters from

$$\{(\theta_o, \phi) : \theta_o \in [0^\circ, 360^\circ), \quad \phi \in [0^\circ, 90^\circ)\} \tag{6.58}$$

to

$$\{(\vartheta_o, \varphi) : \vartheta_o \in [0^\circ, 180^\circ), \quad \varphi \in [0^\circ, 180^\circ)\} \tag{6.59}$$

is essential before using the technique described in Section 6.3.1 and is given as follows:

$$\text{if } \theta_o < 180^\circ \text{ then} \begin{cases} \varphi = \phi \\ \vartheta_o = \theta_o \end{cases} \tag{6.60}$$

$$\text{if } \theta_o \geqslant 180^\circ \text{ then} \begin{cases} \varphi = 180^\circ - \phi \\ \vartheta_o = \theta_o - 180^\circ \end{cases} \tag{6.61}$$

Furthermore, after the construction of the Ambiguous Generator Table of arc lengths, the associated composite directions (azimuth, elevation) can be estimated using Eq. (2.28). Then the (ϑ_o, φ) should be mapped to the original space of the family of the ϕ-curves as follows:

$$\text{if } \varphi < 90^\circ \text{ then} \begin{cases} \phi = \varphi \\ \theta_o = \vartheta_o \end{cases} \tag{6.62}$$

$$\text{if } \varphi \geqslant 90^\circ \text{ then} \begin{cases} \phi = 180^\circ - \varphi \\ \theta_o = 180^\circ + \vartheta_o \end{cases} \tag{6.63}$$

The following example illustrates these additional steps.

Example 6.4 Consider the following planar array of 6 sensors (in half-wavelengths)

$$[\underline{r}_x, \ \underline{r}_y, \ 0] = \begin{bmatrix} -2.5, & -2, & -2.5, & 2.5, & 2, & 2.5 \\ 1.5, & 0, & -1.5, & 1.5, & 0, & -1.5 \\ 0, & 0, & 0, & 0, & 0, & 0 \end{bmatrix}^T \tag{6.64}$$

In order to estimate the ambiguous generator sets associated with the ϕ-curve of constant θ equal to 5° we will use its corresponding composite curve (i.e. φ-curve for $\vartheta_0 = 5^\circ$). This implies that

$$\{(\theta_o, \phi) : \theta_o \in [0^\circ, 360^\circ), \quad \phi \in [0^\circ, 90^\circ)\} \tag{6.65}$$

will be

$$\{(\vartheta_o, \varphi) : \vartheta_o \in [0^\circ, 180^\circ), \quad \varphi \in [0^\circ, 180^\circ)\} \tag{6.66}$$

while the ELA associated with this φ-curve is the following:

$$\underline{r}(\vartheta_0) = [-2.6212, \quad -2.3598, \quad -1.9924, \quad 1.9924, \quad 2.3598, \quad 2.6212]^T \tag{6.67}$$

having a manifold length equal to

$$(l_{\mathrm{m}} \text{ of } \mathcal{A}_{\varphi|5^\circ}) = 35.9943 \tag{6.68}$$

By using Eq. (6.50) in conjunction with the technique described in Section 6.3.1 four AGSs of arc lengths are obtained as shown in Table 6.5. These four AGSs, for this specific ELA of $\vartheta = 5°$, have been constructed from the UBS as shown below:

AGS	UBS
1st	$\underline{s}_{\Delta r_{1,6}}$
2nd	$\underline{s}_{\Delta r_{1,5}}; \underline{s}_{\Delta r_{2,6}}$
3rd	$\underline{s}_{\Delta r_{1,4}}; \underline{s}_{\Delta r_{3,6}}$
4th	$\underline{s}_{\Delta r_{2,4}}; \underline{s}_{\Delta r_{3,5}}$

Note that the directions in degrees associated with the above ambiguous generator sets of arc lengths can be estimated using Eq. (2.28) of Chapter 2 and are given in Table 6.6.

Table 6.5 Ambiguous generator sets (arc lengths).

	s_1	s_2	s_3	s_4	s_5	s_6	ρ_{a}
(1)	0	6.8659	13.7319	20.5978	27.4638	34.3297	5
(2)	0	7.2264	14.4527	21.6791	28.9054	—	4
(3)	0	7.8018	15.6035	23.4053	31.2071	—	4
(4)	0	8.2705	16.5410	24.8114	33.0819	—	4

Table 6.6 Ambiguous generator sets (in degrees) corresponding to Table 6.5.

	(θ_1, ϕ_1)	(θ_2, ϕ_2)	(θ_3, ϕ_3)	(θ_4, ϕ_4)	(θ_5, ϕ_5)	(θ_6, ϕ_6)	ρ_{a}
(1)	$(5°, 0°)$	$(5°, 51.79°)$	$(5°, 76.29°)$	$(185°, 81.69°)$	$(185°, 58.26°)$	$(185°, 24.84°)$	5
(2)	$(5°, 0°)$	$(5°, 53.24°)$	$(5°, 78.64°)$	$(185°, 78.20°)$	$(185°, 52.69°)$	—	4
(3)	$(5°, 0°)$	$(5°, 55.49°)$	$(5°, 82.36°)$	$(185°, 72.51°)$	$(185°, 42.78°)$	—	4
(4)	$(5°, 0°)$	$(5°, 57.29°)$	$(5°, 85.36°)$	$(185°, 67.75°)$	$(185°, 33.05°)$	—	4

6.5.2 *Ambiguities on α-curves/β-curves*

Next let us examine a curve $\mathcal{A}_{\alpha|\beta_o}$ belonging to the family of α-curves $\{\mathcal{A}_{\alpha|\beta_o}, \forall\beta_o : \beta_o \in \Omega_\beta\}$ for a fixed rotation $\Theta = \Theta_o$. In other words let us examine the locus of all manifold vectors over the whole α parameter space $\Omega_\alpha \equiv [\alpha_{\min}, \alpha_{\max}]$ at a particular β_0, i.e.

$$\mathcal{A}_{\alpha|\beta_o} = \{\underline{\mathbf{a}}(\alpha, \beta_o) \in \mathfrak{C}^N, \ \forall\alpha : \alpha \in \Omega_\alpha, \ \beta_o = \text{constant}\} \tag{6.69}$$

with $\alpha_{\min}$ and $\alpha_{\max}$ given by Eq. (5.12). As we have seen in Chapter 5 this family covers and represents the whole of the manifold surface $\mathcal{M}$ of a planar array while all its members are associated with the same ELA

$$\underline{r}(\Psi_0) \quad \text{with} \quad \Psi_0 \triangleq \Theta_o \tag{6.70}$$

but with different "visible" areas specified by the minimum and maximum permissible values of α, i.e. the parameter space is $\Omega_\alpha = [\alpha_{\min}, \alpha_{\max}]$. The same is true for the β-curves but with $\underline{r}(\Theta_o + 90°)$ as their ELA with parameter space $\Omega_\beta = [\beta_{\min}, \beta_{\max}]$.

A few other points must be emphasized here. Firstly, from Eq. (6.3), the following relation can be derived:

$$\underline{r}(\Theta) = -\underline{r}(\Theta + 180°) \tag{6.71}$$

which implies that for a given rotation Θ_o of the x–y frame and a given β_o, the resulting α-curve $\mathcal{A}_{\alpha|\beta_o}(\Theta_o)$ has the same length as the α-curve $\mathcal{A}_{\alpha|\beta_o}(\Theta_o + 180°)$ resulting from rotation $\Theta_o + 180°$. This is the reason why only frame rotations Θ_o in the region $[0°, 180°]$ should be considered. Secondly, as already mentioned, the roles of α and β can be interchanged by simply replacing the rotation Θ_o of the x–y frame with $\Theta_o + 90°$ and therefore the β-curves are effectively transposed versions of the α-curves. This means that the α-curves for some rotation Θ_o, are the same as the β-curves for rotation $\Theta_o + 90°$. This is the reason the β-curves are neglected throughout this analysis and only the α-curves are considered.

Based on the above discussion the ambiguous generator sets of arc lengths of the α-curve of $\beta_o = 90°$, i.e. $\mathcal{A}_{\alpha|90°}$, can be estimated by partitioning this curve according to the elements (arc lengths) of Eq. (6.50) using $\Psi_0 \triangleq \Theta_o$.

Note that the α cone-angle parameter, corresponding to the arc lengths s_k of the α-curve, can be calculated by the following equation:

$$\alpha_k = \arccos\left(\sin\beta_o - \frac{s_k}{\pi\|\underline{r}(\Theta_o)\|}\right) \tag{6.72}$$

and that the α-curve with $\beta_o = 90°$ for a frame rotation Θ_o has identical AGSs with the composite ϕ-curve associated with the composite azimuth $\vartheta_o = \Theta_o$.

Indeed if $\beta_o = 90°$ then

$$\theta_o = \begin{cases} \Theta_o & \text{if } \alpha \leq 90° \\ \Theta_o + 180° & \text{if } \alpha > 90° \end{cases} \tag{6.73}$$

$$\phi = \begin{cases} \alpha & \text{if } \alpha \leq 90° \\ 180° - \alpha & \text{if } \alpha > 90° \end{cases} \tag{6.74}$$

Now, based on Eq. (6.50), let us assume that all AGSs of arc lengths of the α-curve with $\beta_o = 90°$, have been constructed. Then consider another α-curve of β_o different to 90°. For this second curve let $s_{\min}$ and $s_{\max}$ be the arc lengths corresponding to the parameter angles $\alpha_{\min}$ and $\alpha_{\max}$ (given by Eq. (5.12)), respectively. This curve has ambiguous sets but not ambiguous generator sets, since $s_{\min} \neq 0$ for any $\beta \neq 90°$, which implies that the first condition of the definition of the ambiguous generator set (see Definition 6.5) is not satisfied. Let us define the ambiguous sets with their first element equal to $s_{\min}$ as the *first permissible ambiguous sets.* These can be found by rotating all the ambiguous generator sets of the α-curve with $\beta_o = 90°$ by $s_{\min}$, subject to the condition that the maximum element of each set cannot exceed $s_{\max}$. Thus it is clear that the ambiguous generator sets for the whole family of α-curves, for a given frame rotation Θ_o, can be provided by examining only the α-curve of $\beta_o = 90°$. The ambiguities of any other α-curve can be generated by a simple rotation (Theorem 6.1) of those ambiguous generator sets, making sure that its maximum element s_c is smaller than the manifold length l_m of $\mathcal{A}_{\alpha|\beta_o}$.

Based on the above discussion it is clear that the Ambiguous Generator Table in arc lengths should be constructed for the α-curve with $\beta_o = 90°$, using the technique of Section 6.3.1. Then, for a specific α-curve with $\beta_o < 90°$ only the AGSs should be kept whose last element s_c is smaller than the length of the α-curve given by $l_{m_{\alpha|\beta_o}} = 2\pi\|\underline{r}(\Theta_o)\| \sin\beta_o$. These sets can be expressed from arc lengths to degrees using Eq. (6.72) but using an additional term $s_{\min}$ which takes into account the fact that for $\beta_o \neq 90°$ the minimum value of α is not 0°. That is

$$\alpha_k = \arccos\left(1 - \frac{s_k + \overbrace{\pi\|\underline{r}(\Theta_o)\|(1 - \sin\beta_o)}^{s_{\min}}}{\pi\|\underline{r}(\Theta_o)\|}\right) \tag{6.75}$$

Note that $s = 0$ must correspond to $\alpha_k = \arccos(\sin\beta_o) = 90° - \beta_o, \forall k$.

Furthermore, some of these sets may also be AGSs of various other α-curves with $\beta_o \neq 90°$. However the AGS of arc lengths for an α-curve with $\beta_o \neq 90°$, although numerically identical to those of an α-curve with $\beta_o = 90°$ (or equivalently to those of ϕ-curve $\vartheta_o = \Theta_o$), correspond to different sets of directions having different azimuth and elevation angles. A number of these properties are shown in the following example.

Example 6.5 Consider a planar array whose sensor locations (in half-wavelengths) are given below:

$$[\underline{r}_x, \ \underline{r}_y, \ 0] = \begin{bmatrix} -2.6, & -1.3, & 1.3, & 2.6 \\ 2, & -0.5, & -1.5, & 0 \\ 0, & 0, & 0, & 0 \end{bmatrix}^T \tag{6.76}$$

The Equivalent Linear Array of the family of α-curves $\{\mathcal{A}_{\alpha|\beta_o}, \ \forall \beta_o : \beta_o \in \Omega_\beta\}$ for a fixed rotation of the x-y frame $\Theta_o = 0°$ is (see Eq. (6.3))

$$\underline{r}(\Theta_o) = \underline{r}_x \cos\Theta_o + \underline{r}_y \sin\Theta_o = \begin{bmatrix} -2.6 \\ -1.3 \\ 1.3 \\ 2.6 \end{bmatrix} \tag{6.77}$$

The manifold length of this ELA $\underline{r}(\Theta_o)$ is $l_m = 2\pi\|\underline{r}(\Theta_o)\| = 25.8299$. By employing the technique of Section 6.3.1, Table 6.7 of the AGSs can be obtained where the relation between AGSs and UBSs is given in Table 6.8.

Table 6.7 AGSs (arc lengths) of the ELA of Eq. (6.77).

	s_1	s_2	s_3	s_4	p_a
(1)	0	19.8692	—	—	1
(2)	0	6.6231	13.2462	—	2
(3)	0	4.9673	24.8365	—	2
(4)	0	9.9346	14.9019	24.8365	3
(5)	0	4.9673	9.9346	14.9019	3

Table 6.8 UBS associated with the AGSs of Table 6.7.

AGS	UBS
1st	$\underline{s}_{\Delta r_{1,2}}; \underline{s}_{\Delta r_{2,3}}; \underline{s}_{\Delta r_{2,4}}; \underline{s}_{\Delta r_{1,3}}; \underline{s}_{\Delta r_{1,4}}; \underline{s}_{\Delta r_{3,4}}$
2nd	$\underline{s}_{\Delta r_{1,3}}; \underline{s}_{\Delta r_{2,4}}$
3rd to 5th	$\underline{s}_{\Delta r_{1,4}}$

The sets of Table 6.7 are the ambiguous generator sets for the whole family of α-curves (since ambiguities of any α-curve with $\beta_o < 90°$ will relate to a subset of the above AGSs). To see this property let us consider $\beta_o = 90°$. In this case, all the AGSs of Table 6.7 are certainly included since the length l_m of this α-curve $\mathcal{A}_{\alpha|90°}$ is equal to the manifold length of the ELA $l_m = 25.8299$. The values in Table 6.7, transformed in degrees, are shown in Table 6.9. If a different α-curve of $\beta_o < 90°$ is chosen then its length is

$$l_m = 2\pi \|\underline{r}(\Theta_o)\| \sin \beta_o = 25.8299 \sin \beta_o \tag{6.78}$$

In this case only the rows in Table 6.7 with the last element less than l_m should be kept. Indeed for the α-curve with $\beta_o = 74°$, i.e. $\mathcal{A}_{\alpha|74°}$, then $l_m = 24.8293$ and therefore the AGSs will be as shown in Table 6.10.

The α-curve $\mathcal{A}_{\alpha|50°}$, having a length $l_m = 19.7869$ will be associated with only two out of the five AGS of the family. These are the ones with their last element smaller than, or equal to, the length of the curve 19.7869 (i.e. $\underline{s}_{\Delta r_{1,3}}$ and $\underline{s}_{\Delta r_{1,3}}$) and are given in Table 6.11.

Table 6.9 AGSs (in degrees) of the ELA of Eq. (6.77).

	(α_1, β_o) or (ϕ_1, θ_o)	(α_2, β_o) or (ϕ_2, θ_o)	(α_3, β_o) or (ϕ_3, θ_o)	(α_4, β_o) or (ϕ_4, θ_o)	ρ_a
(1)	$(0°, 90°)$	$(122.58°, 90°)$	—	—	1
(2)	$(0°, 90°)$	$(60.84°, 90°)$	$(91.47°, 90°)$	—	2
(3)	$(0°, 90°)$	$(52.02°, 90°)$	$(157.38°, 90°)$	—	2
(4)	$(0°, 90°)$	$(76.65°, 90°)$	$(98.85°, 90°)$	$(157.38°, 90°)$	3
(5)	$(0°, 90°)$	$(52.02°, 90°)$	$(76.65°, 90°)$	$(98.85°, 90°)$	3

Table 6.10 AGSs of the ELA of Eq. (6.77) for $\beta_o = 74°$.

	(α_1, β_o)	(α_2, β_o)	(α_3, β_o)	(α_4, β_o)	ρ_a
(1)	$(16°, 74°)$	$(125.25°, 74°)$	—	—	1
(2)	$(16°, 74°)$	$(63.36°, 74°)$	$(93.69°, 74°)$	—	2
(5)	$(16°, 74°)$	$(54.79°, 74°)$	$(78.93°, 74°)$	$(101.10°, 74°)$	3
	(ϕ_1, θ_1)	(ϕ_2, θ_2)	(ϕ_3, θ_3)	(ϕ_4, θ_4)	ρ_a
(1)	$(16°, 0°)$	$(154.47°, 50.24°)$	—	—	1
(2)	$(16°, 0°)$	$(31.58°, 58.24°)$	$(103.15°, 73.56°)$	—	2
(5)	$(16°, 0°)$	$(25.55°, 50.27°)$	$(55.14°, 70.37°)$	$(124.94°, 70.35°)$	3

Table 6.11 AGSs of the ELA of Eq. (6.77) for $\beta_o = 50°$.

	(α_1, β_o)	(α_2, β_o)	(α_3, β_o)	(α_4, β_o)	ρ_a
(2)	$(40°, 50°)$	$(75.33°, 50°)$	$(105.05°, 50°)$	—	2
(5)	$(40°, 50°)$	$(67.58°, 50°)$	$(90.18°, 50°)$	$(112.82°, 50°)$	3
	(ϕ_1, θ_1)	(ϕ_2, θ_2)	(ϕ_3, θ_3)	(ϕ_4, θ_4)	ρ_a
(2)	$(40°, 0°)$	$(68.50°, 46.30°)$	$(112°, 46.11°)$	—	2
(5)	$(40°, 0°)$	$(59.32°, 41.63°)$	$(90.28°, 50°)$	$(121.10°, 41.35°)$	3

Table 6.12 AGSs of the ELA of Eq. (6.77) for $\beta_o = 35°$.

	(α_1, β_o)	(α_2, β_o)	(α_3, β_o)	(α_4, β_o)	ρ_a
(2)	$(55°, 35°)$	$(86.52°, 35°)$	$(116.88°, 35°)$	—	2
	(ϕ_1, θ_o)	(ϕ_2, θ_o)	(ϕ_3, θ_o)	(ϕ_4, θ_o)	ρ_a
(2)	$(55°, 0°)$	$(85.76°, 34.77°)$	$(118.89°, 20.67°)$	—	2

Finally, Table 6.12 gives the only AGS associated with the α-curve $\mathcal{A}_{\alpha|35°}$ (curve length $l_m = 14.8154$), while no ambiguity can be identified for any α-curve with $\beta_o < 30°$ or $\beta_o > 150°$.

All these are shown in Fig. 6.4 which plots the length of the α-curves versus the value of β_o where the darker the shade of the area, the larger the number of ambiguous sets that can be identified.

Note that in order to have a set of M (azimuth, elevation) directions belonging to the same α-curve, there should be a frame rotation $\Theta = \Theta_o$ such that

$$\arccos(\cos \underline{\phi} \odot \sin(\underline{\theta} - \Theta \underline{1}_M)) = k \underline{1}_M \tag{6.79}$$

where k is a constant. Then these directions which satisfy Eq. (6.79) belong to the same α-curve $\mathcal{A}_{\alpha|\beta_o}$, i.e correspond to a common $\beta_o = k$ for this rotation Θ_o. For instance this can be seen for the following planar array:

$$[\underline{r}_x, \ \underline{r}_y, \ \underline{0}] = \begin{bmatrix} -1, & 2, & -4, & 3 \\ 0, & 0.5, & 1.5, & -2 \\ 0, & 0, & 0, & 0 \end{bmatrix}^T \tag{6.80}$$

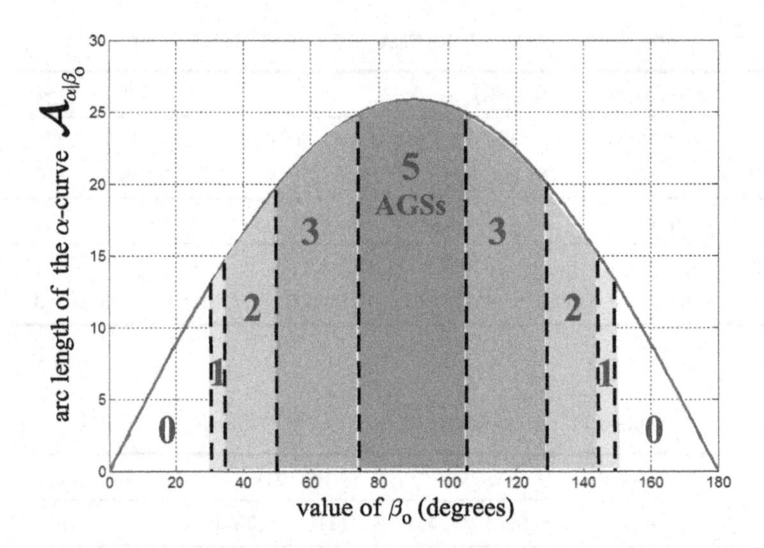

Fig. 6.4 Manifold length for the family of α-curves. In the region of the darker shade the number of AGS is 5 while no AGS can be identifed in the white region (0 AGSs).

operating in the presence of four signals with directions:

	1st DOA	2nd DOA	3rd DOA	4th DOA
$\underline{\theta}$	47°	54.72°	129.57°	203.36
$\underline{\phi}$	0°	59.09°	82°	58.91°

(6.81)

Indeed in this case Eq. (6.79) is satisfied for $\Theta = \Theta_o = 39°$ with constant $k = 82°(= \beta_o)$ as can be seen from Fig. 6.5.

This indicates that these directions are on the same α-curve which is the manifold of the ELA $\underline{r}(\Psi_0)$ with $\Psi_0 = \Theta_o$, i.e.

$$\underline{r}(\Theta_o) = \underline{r}_x \cos\Theta_o + \underline{r}_y \sin\Theta_o \tag{6.82}$$

This α-curve (manifold) can be found using Eqs. (5.3) and (5.4) to transform the set of $(\underline{\theta}, \underline{\phi})$ to cone-angles for the frame rotation Θ_o. Thus for $\Theta_o = 39°$ the directions $(\underline{\theta}, \underline{\phi})$ are transformed to cone-angles $(\underline{\alpha}, \underline{\beta})$ given

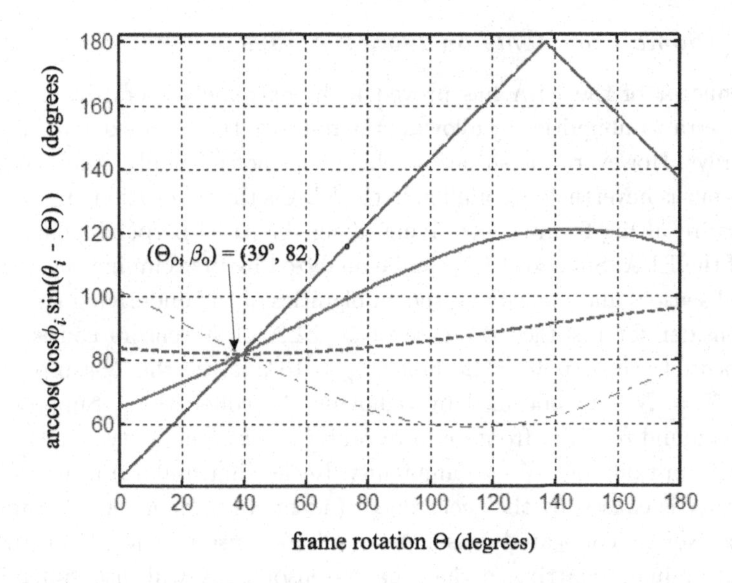

Fig. 6.5 Plot of the four equations given by Eq. (6.79) for different rotation Θ.

in the following table:

	1st DOA	2nd DOA	3rd DOA	4th DOA
$\underline{\alpha}$	8°	60.37°	90.08°	119.82°
$\underline{\beta}$	82°	82°	82°	82°

This implies that the common β_o is 82° for this rotation Θ_o (i.e. belong to the same α-curve $\mathcal{A}_{\alpha|\beta_o}$). This curve is the manifold of the following ELA $\underline{r}(\Psi_0)$ with $\Psi_0 = \Theta_o$:

$$\underline{r}(39°) = \underline{r}_x \cos 39° + \underline{r}_y \sin 39° = \left[-2.1646, -0.7771, 1.0728, 1.8690\right]^T \tag{6.83}$$

Note that if there exists an intersensor difference $\Delta r_{ij} = |r_i(\Theta_o) - r_j(\Theta_o)|$ for which all elements of $\underline{\alpha}$ satisfy

$$\underline{\alpha} = \arccos\left(\sin\beta_0 \underline{1}_M - \frac{2}{\Delta r_{ij}}[1, 2, \ldots, M]^T\right) \tag{6.84}$$

then this set is an ambiguous set.

6.5.3 *Some Comments on Planar Arrays*

The concept of the ELA has proved to be extremely useful in analyzing planar array ambiguities by allowing the re-use of the results derived for linear arrays. However, one situation which does not normally occur in linear arrays but is nevertheless common in the ELA is the collocation of projected sensors, resulting from sensors lying along the same perpendicular to the line of the ELA. Such an ELA can be analyzed by first eliminating the collocated sensors and identifying the ambiguities of the underlying structure.

Consider for instance the case of an ELA of N-sensors consisting of ℓ collocated sensor pairs. The first step is to analyze the $\widetilde{N}$-sensor array (with $\widetilde{N} = N - \ell$) obtained by removing the collocations. Suppose this array is found to suffer from uniform rank-$(\widetilde{N} - k)$ ambiguity, i.e. the ELA suffers from rank-$(N - \ell - k)$ ambiguity. Recall that uniform rank-$(\widetilde{N} - k)$ ambiguity is caused by the occurrence of identical rows in the determinant matrix. Now, ℓ collocated sensor pairs will give rise to ℓ identical rows in the determinant matrix, so that the N-sensor ELA will also suffer from rank-$(\widetilde{N} - k)$ or, equivalently, rank-$(N - \ell - k)$ ambiguity.

Note that an ELA has two or more collocated sensors when for one or more different pairs of sensors the parameter Ψ_0 satisfies the following equation:

$$\tan \Psi_0 = -\frac{r_{xi} - r_{xj}}{r_{yi} - r_{yj}}, \quad i \neq j \tag{6.85}$$

where Ψ_0 is given by Eq. (6.4), $\Psi_0 \in [0, 180°)$ and (r_{xi}, r_{yi}), (r_{xj}, r_{yj}) represent the locations of the ith and j^{th} sensors. It can be proven that, for a given array geometry, the maximum number $n_{\max}$ of ELAs with collocated sensors, is given by:

$$n_{\max}(\mathbf{r}) = \frac{N^2 - N}{2} - 2\binom{m}{2}\delta_1 - 2m\delta_2 \tag{6.86}$$

with m being the number of symmetric pairs with respect to the origin in the planar array,

$$\delta_1 = \begin{cases} 1 & \text{if } m > 1 \\ 0 & \text{otherwise} \end{cases} \tag{6.87}$$

and

$$\delta_2 = \begin{cases} 1 & \text{if the array has a sensor at the origin} \\ 0 & \text{otherwise} \end{cases} \tag{6.88}$$

To better illustrate the condition of collocated sensors, consider for instance a uniform circular array (UCA) with 8 elements and 1.5 half-wavelengths intersensor spacing. Equation (6.86) will give us that the maximum number of ELAs with collocated sensors is $n_{\max} = 8$. Indeed the array has eight ELAs with collocated sensors corresponding to $\Psi_0 = 0°$, 22.5°, 45°, 67.5°, 90°, 112.5°, 135° and 157.5°. In Fig. 6.6, the ELA for $\Psi_0 = 45°$ is presented (with the circles) and has only one ambiguous generator set of rank two. Furthermore in this figure the number of ambiguous generator sets over the whole parameter space are shown grouped by rank. The above example is a characteristic one which illustrates the importance of the array geometry in the ambiguity problem and in the presence of lower rank ambiguities. Note that all symmetric array geometries (grid, X, Y shaped arrays, etc.) have similar properties to those of the uniform circular array.

Within this new framework, the results reported so far concerning ambiguities present in linear array manifolds are directly applicable to the p-curves of planar array manifolds via the concept of the equivalent linear

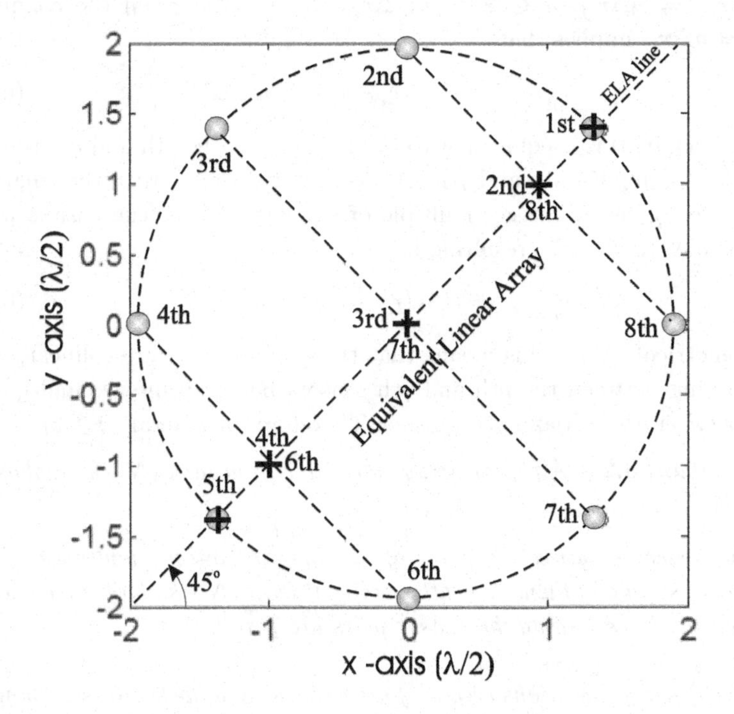

Fig. 6.6 ELA with co-located sensors.

array. Based on similar arguments to the linear array case the following statements about ambiguities inherent in planar arrays can be made:

Proposition 6.1

(a) A planar array is bound to suffer from rank-$(N-1)$ ambiguity if two of its sensors are at least $(N-1)$ half-wavelengths apart.

(b) Furthermore, rank-$(N-k)$ ambiguity will exist if there exists an ELA, where k intersensor spacings are integer multiples of the smallest among them, and the latter is at least $(N-k)$.

Let us now investigate the implications of condition (b) on the planar array configuration,

$$\begin{aligned}\mathbf{r} &= [\underline{r}_x, \underline{r}_y, \underline{0}_N] \\ &= [\underline{r}_1, \underline{r}_2, \ldots, \underline{r}_N]^T\end{aligned} \tag{6.89}$$

where $\underline{r}_i = [x_i, y_i, 0]^T$ is the position of the ith sensor in half-wavelengths. Consider the ELA $\underline{r}(\Psi_0) \triangleq \underline{r}_x \cos\Psi_0 + \underline{r}_y \sin\Psi_0$ for the case when $k=2$, say $\Delta r_{pq} = n\Delta r_{ij}$, $n \in \mathcal{Z}^+$ and $\Delta r_{ij} \geqslant (N-2)$. Then the condition $\Delta r_{pq} = n\Delta r_{ij}$ implies that

$$\underline{r}_p - \underline{r}_q = \pm n\left(\underline{r}_i - \underline{r}_j\right), \quad n \in \mathcal{Z}^+ \tag{6.90}$$

Geometrically, this is equivalent to the line joining the pth and qth sensors and that joining the ith and jth sensors, being parallel, with the length of the former being an integer multiple of the latter. A special case is when $\Delta r_{pq} = n\Delta r_{qj}; n \in \mathcal{Z}^+$, resulting in

$$\underline{r}_p - \underline{r}_q = \pm n\left(\underline{r}_q - \underline{r}_j\right), \quad n \in \mathcal{Z}^+ \tag{6.91}$$

In geometrical terms, this means that the three sensors are colinear, with the spacing between the pth and qth sensors being an integer multiple of that between the ith and jth sensors. Therefore, in general $(k \geqslant 2)$.

Proposition 6.2 *A planar array will suffer from rank-$(N-k)$ ambiguity if either*

(a) the spacings between k pairs of sensors are integer multiples of the smallest among them, the latter being at least $(N-k)$ half-wavelengths, and the lines joining the sensor pairs are parallel,

or

(b) $(k+1)$ sensor locations (vectors) are colinear and the k intersensor spacings between adjacent sensor pairs are integer multiples of the smallest among them, with the latter being at least $(N-k)$ half-wavelengths.

Note that the same factors which contribute to the growth of the number of ambiguous generator sets in linear arrays are applicable to planar arrays via the equivalent linear array.

Corollary 6.6 *The causes of ambiguities in planar arrays are namely the colinearity of the sensor locations, uniformity in the sensor spacings and the size of the common spacing.*

6.5.4 *Ambiguous Generator Lines*

We have seen that for a given ELA $\underline{r}(\Psi_0)$ the AGSs can be constructed using the technique presented in Section 6.3.1. It is also evident from the previous sections that in a planar array, the whole family of α-curves is represented with only one ELA, with Ψ_0 being the frame rotation Θ_o, i.e.

$$\boxed{\text{ELA of the family of } \alpha\text{-curves} = \underline{r}(\Theta_o)} \tag{6.92}$$

while an infinite number of ELAs are associated with the family of ϕ-curves

$$\boxed{\text{set of ELAs of the family of } \phi\text{-curves} = \left\{\underline{r}(\theta_o) \in \mathcal{R}^N,\ \ \forall\theta_o \in \Omega_\theta\right\}} \tag{6.93}$$

By varying Ψ_0 (θ_o or Θ_o) over its parameter space $\Omega_{\Psi_0} = [0°,\ 180°)$ we will get the set of *ambiguous generator lines* of the whole family of p-curves to which the elements of the ambiguous generator sets belong.

Example 6.6 Let us consider the array of Example 6.4. The first ambiguous generator set of this array, for this specific ELA of $\vartheta = 5°$, has rank of ambiguity equal to 5 and is given below:

$$\text{AGS} = [0, 6.8659, 13.7319, 20.5978, 27.4638, 34.3297]$$

while its elements are shown as dots in Fig. 6.7. Furthermore, in the same figure the locus of the manifold lengths of all ELAs, i.e. $\{l_m(\Psi_0),\ \forall\Psi_0\}$ is also shown.

By varying Ψ_0 over its parameter space $\Omega_{\Psi_0} = [0°, 180°)$ we will get the set of *ambiguous generator lines* of the ϕ-curves to which the elements of the ambiguous generator sets of arc lengths belong.

In these lines the AGS (in arc length) are associated with all ELA $\underline{r}(\Psi_0), \forall\Psi_0$ with $\Psi_0 = \Theta_o$ or ϑ_o. This implies that, for variable Θ_o in the region 0° to 180°, and for the array of Eq. (6.64) the ambiguous generator lines of all families of α-curves are numerically, in arc lengths, equal to those of the family of ϕ-curves. Thus for the array of Eq. (6.64),

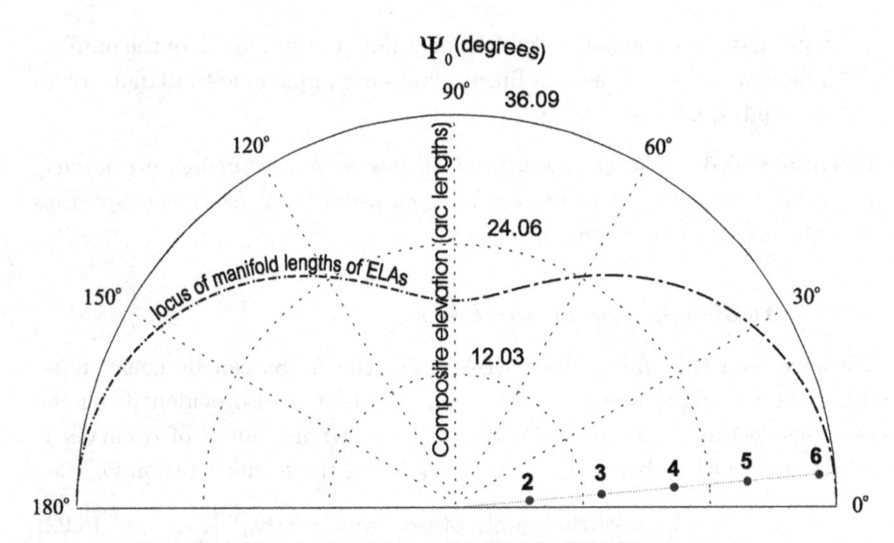

Fig. 6.7 The locus of the manifold lengths of ELA of Example 6.4. The dots represent the 1st AGS (rank 5) given in Table 6.5.

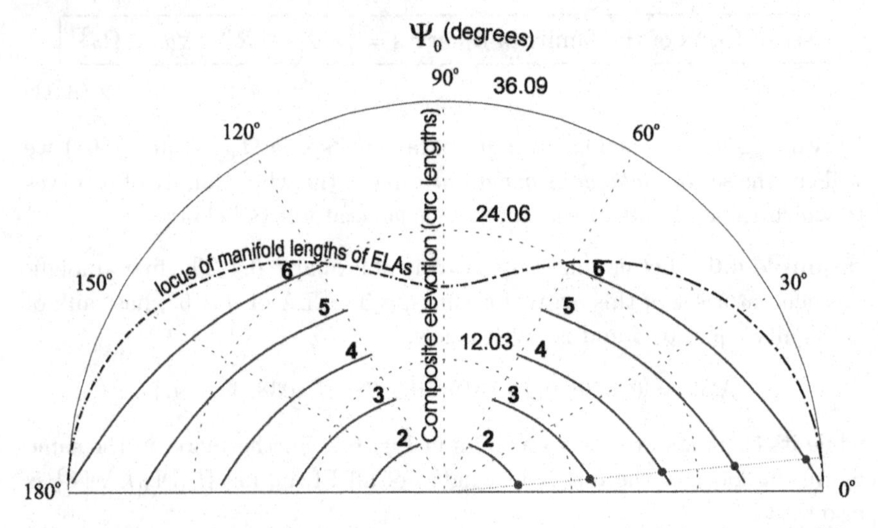

Fig. 6.8 The set of ambiguous generator lines of rank 5 in which the elements of the 1st AGS of Table 6.5 (Example 6.4) belong.

Figs. 6.8–6.11 not only represent the set of ambiguous generator lines of the ϕ-curves but also of the α and β-curves by generalizing the polar axes to (Ψ_0, s).

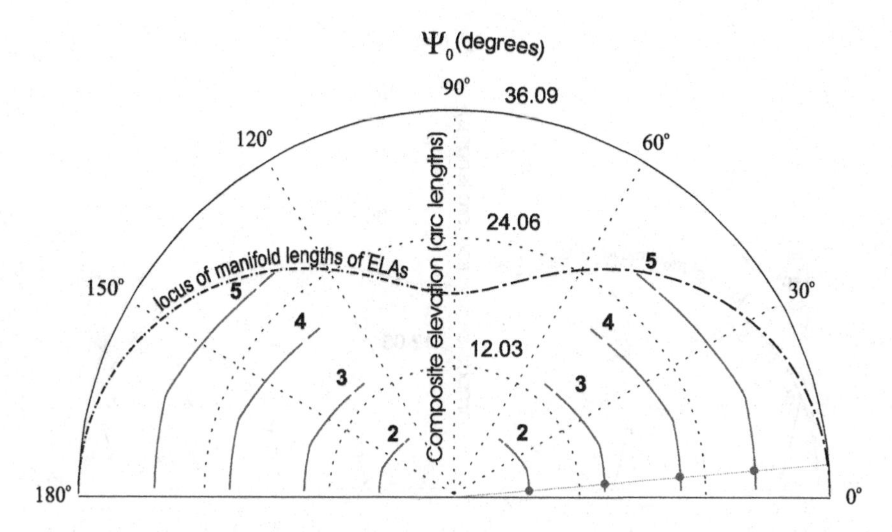

Fig. 6.9 The set of ambiguous generator lines of rank 4 in which the elements of the 2nd AGS of Table 6.5 (Example 6.4) belong.

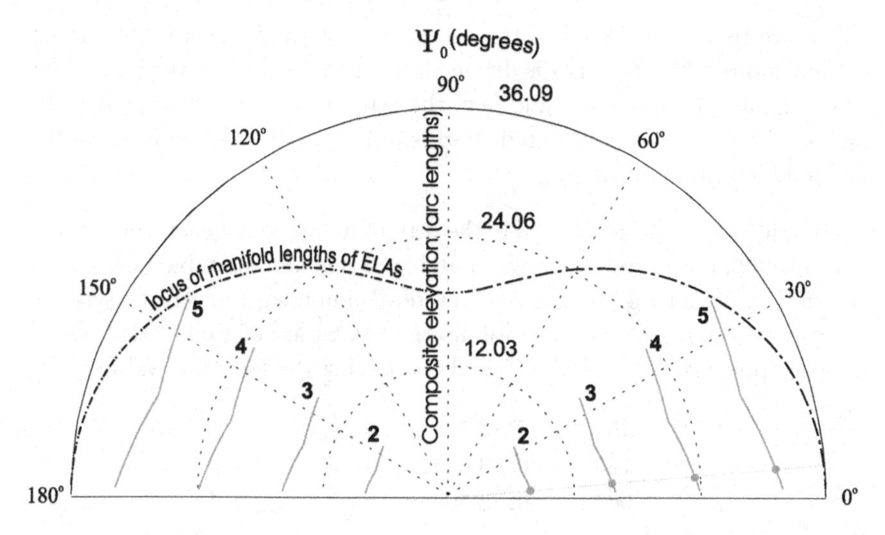

Fig. 6.10 The set of ambiguous generator lines of rank 4 in which the elements of the 3rd AGS of Table 6.5 (Example 6.4) belong.

Thus, Fig. 6.8 shows the set of *ambiguous generator lines* of rank 5 in which the first row of Table 6.5 belongs. As it can be seen in this figure, the set of *ambiguous generator lines* ceases to exist at $\widehat{\theta}_o = 0$ and for values of $\widehat{\theta}_o$ in the region 61° to 119° while for $\widehat{\theta}_o$ in the region 119° to 180° its

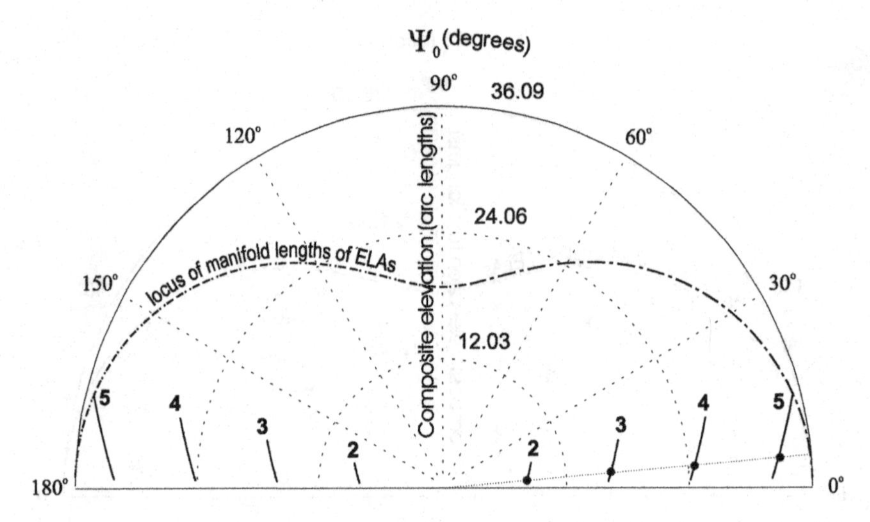

Fig. 6.11 The set of ambiguous generator lines of rank 4 in which the elements of the 4th AGS of Table 6.5 (Example 6.4) belong.

values are the mirror image of the values from 0° to 61°. Note that the set of ambiguous generator lines is defined only in those areas of the parameter space $\Omega_{\widehat{\theta}}$, at which the last line (e.g. the 6th line in Fig. 6.8) in which the largest element of the associated AGS is located, is below the locus of the manifold length of all ELAs.

In addition, in Figs. 6.9–6.11, the sets of ambiguous generator lines in which the 2nd, 3rd and 4th rows of ambiguous generator table belong, are shown. Finally, in Fig. 6.12 three "discrete" ambiguous generator sets are shown as "squares." Two of these discrete AGSs are of rank 2 and one of rank 1, appear at $\widehat{\theta}_o = 45°$, 135° and 90° having the following values:

$$\begin{aligned}
\widehat{\theta}_o = 45° : &\quad [0, 13.5732, 27.465]; &\quad \rho_a = 2\\
\widehat{\theta}_o = 135° : &\quad [0, 13.5732, 27.1465]; &\quad \rho_a = 2\\
\widehat{\theta}_o = 90° : &\quad [0, 12.5664]; &\quad \rho_a = 1
\end{aligned}$$

6.6 Ambiguities and Manifold Length

It must be apparent by now that the existence of ambiguities is intimately related to the topology of the array manifold and especially to the length of the manifold curve. Intuitively, longer manifolds are more prone to ambiguous situations. Consider the set of arc lengths resulting from the uniform partitions of the manifold length l_m by the spacing between the ith and jth

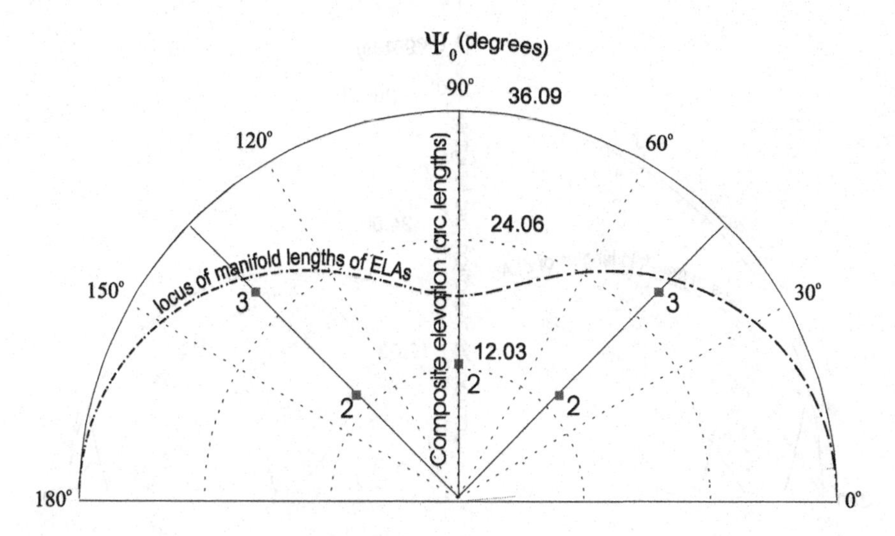

Fig. 6.12 "Discrete" AGSs of corresponding to ELA of 45°, 90° and 135° for the array of the Example 6.4.

sensors of an N-sensor linear array:

$$s = \left[0, \frac{l_m}{\Delta r_{ij}}, \frac{2l_m}{\Delta r_{ij}}, \ldots, \frac{(c-1)\, l_m}{\Delta r_{ij}}\right] \tag{6.94}$$

Now $l_m = 2\pi\|\underline{r}\|$ and hence the set s can be rewritten as

$$s_{\Delta\tilde{r}_{ij}} = \left[0, \frac{2\pi}{\Delta\tilde{r}_{ij}}, \frac{4\pi}{\Delta\tilde{r}_{ij}}, \ldots, \frac{2\pi(c-1)}{\Delta\tilde{r}_{ij}}\right] \tag{6.95}$$

where

$$\Delta\tilde{r}_{ij} = |\tilde{r}_i - \tilde{r}_j| \quad \text{with} \quad \underline{\tilde{r}}_i \triangleq \frac{r_i}{\|\underline{r}\|}$$

For $s_{\Delta\tilde{r}_{ij}}$ to be a rank-$(N-1)$ ambiguous generator set, it must consist of at least N terms and therefore

$$l_m \geqslant \frac{2(N-1)\pi}{\Delta\tilde{r}_{ij}} \tag{6.96}$$

Considering the largest value of $\Delta\tilde{r}_{ij}$, a sufficient condition for the N-sensor array to suffer from rank-$(N-1)$ ambiguity is then given by

$$l_m \geqslant \frac{2\,(N-1)\,\pi}{\Delta\tilde{r}_{1,N}} \tag{6.97}$$

Recall that the eigenvalues of the Cartan matrix are equivalent to j times the normalized sensor locations of the corresponding array and mirror array.

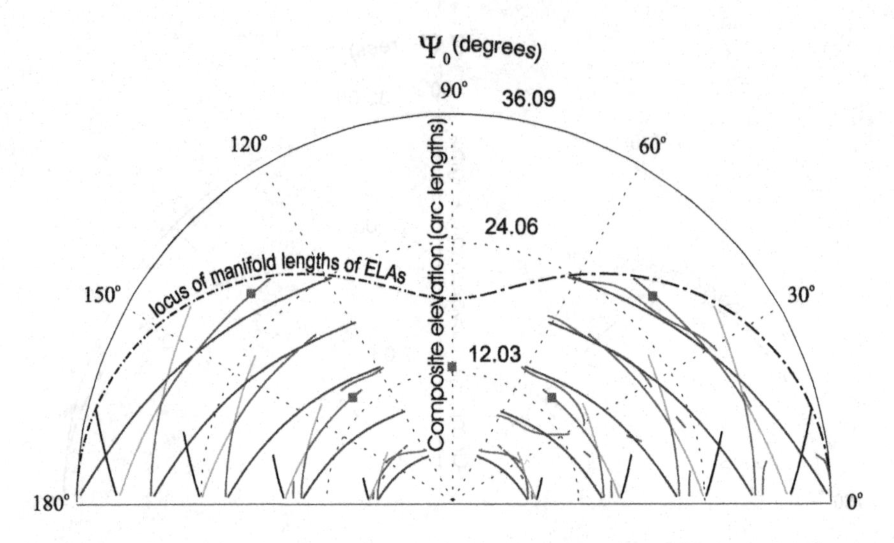

Fig. 6.13 All the sets of ambiguous generator lines of the array of Example 6.4 shown together with the "discrete" AGSs.

In the case of a symmetric linear array, the normalized sensor locations are simply given by j times the eigenvalues of the corresponding Cartan matrix (since the array and its mirror image about the centroid are identical). Therefore, in the symmetric linear array case, the denominator $\Delta\tilde{r}_{1,N}$ in Eq. (6.97) is equal to the eigenvalue spread of the Cartan matrix-spread($\mathbb{C}$). Note that the spread of a matrix $\mathbb{A}$ with eigenvalues denoted by λ, is defined as:

$$\text{spread}\,(\mathbb{A}) \triangleq \max_{i,j} |\lambda_i - \lambda_j| \tag{6.98}$$

and it satisfies the following inequalities [22]:

$$\text{spread}(\mathbb{A}) \leqslant \sqrt{2}\|\mathbb{A}\|_F \tag{6.99}$$

$$\text{spread}(\mathbb{A}) \geqslant \max_{i,j}\left(|a_{ij}| + |a_{ji}|\right), \quad \text{if } \mathbb{A} \text{ is normal} \tag{6.100}$$

Recall that the Cartan matrix is a real normal matrix ($\mathbb{C}\mathbb{C}^T = \mathbb{C}^T\mathbb{C}$), which in the case of a symmetric array, is given by

$$\mathbb{C} = \begin{bmatrix} 0 & -\kappa_1 & 0 & \dots & 0 \\ \kappa_1 & 0 & -\kappa_2 & \dots & 0 \\ 0 & \kappa_2 & 0 & \dots & 0 \\ \vdots & \vdots & \vdots & \ddots & \vdots \\ 0 & 0 & 0 & \kappa_{N-1} & 0 \end{bmatrix} \tag{6.101}$$

where κ_1, known as the first or principal curvature, is the largest of all the curvatures. Also, from Eq. (2.76) (page 52), $\|\mathbb{C}\|_F = 1$ in the symmetric array case. Hence, from Eq. (6.100), the spread of the Cartan matrix satisfies the following inequality:

$$2\kappa_1 \leq \text{spread}\,(\mathbb{C}) \leq \sqrt{2} \tag{6.102}$$

or equivalently,

$$2\kappa_1 \leq \Delta\tilde{r}_{1,N} \leq \sqrt{2} \tag{6.103}$$

Then from Eq. (6.97), it can be deduced that:

Proposition 6.3 *If the length l_{m} and first curvature κ_1 of the manifold of an N-sensor symmetric LA or ELA satisfy the following condition:*

$$l_{\mathrm{m}} \geqslant \frac{(N-1)\,\pi}{\kappa_1} \tag{6.104}$$

then, rank-$(N-1)$ ambiguity, resulting from uniform partitions of the manifold, is bound to exist. If instead,

$$l_{\mathrm{m}} < \sqrt{2}\,(N-1)\,\pi \tag{6.105}$$

then, uniform rank-$(N-1)$ ambiguity cannot exist along the manifold.

It is however important to note that the condition set out in Eq. (6.105) does not guarantee the absence of other types of rank-$(N-1)$ ambiguity, for instance, those corresponding to non-uniform partitions of the array manifold.

6.7 Appendices

6.7.1 *Proof of Theorem 6.1*

Since

$$\underline{s} = [s_1, s_2, \ldots, s_c]^T \tag{6.106}$$

is an ambiguous set the matrix $\mathbb{A}(\underline{s}) = [\underline{\mathbf{a}}(s_1), \underline{\mathbf{a}}(s_2), \ldots, \underline{\mathbf{a}}(s_c)] \in \mathfrak{C}^{N\times c}$ is rank deficient. This means that any submatrix $\mathbb{A}_c(\underline{s})$ of $\mathbb{A}(\underline{s})$ which has exactly c rows, is singular. That is

$$\det(\mathbb{A}_c(\underline{s})) = 0 \tag{6.107}$$

Consider now the set

$$\begin{aligned}\underline{\check{s}} &= [s_1 + \Delta s, s_2 + \Delta s, \ldots, s_c + \Delta s]^T \\ &= \underline{s} + \Delta s \cdot \underline{1}_c \end{aligned} \tag{6.108}$$

The manifold vector corresponding to the i-th element of $\underline{\check{s}}$ is

$$\underline{\mathbf{a}}(s_i + \Delta s) = \underline{\mathbf{a}}(s_i) \odot \underline{\mathbf{a}}(\Delta s) \tag{6.109}$$

Therefore, the matrix $\check{\mathbb{A}}$ with columns the manifold vectors corresponding to the set $\underline{\check{s}}$ is

$$\check{\mathbb{A}} = \mathbb{A}(\underline{\check{s}}) = \mathbb{A}(\underline{s} + \Delta s \underline{1}_c) = \mathbb{A}(\underline{s}) \odot \mathbb{A}(\Delta s \underline{1}_c) \tag{6.110}$$

The determinant of any submatrix $\check{\mathbb{A}}_c$ of $\check{\mathbb{A}}$ which has exactly c rows, is

$$\begin{aligned}\mathbf{det}(\check{\mathbb{A}}_c) &= \mathbf{det}(\mathbb{A}_c(\underline{s}) \odot \mathbb{A}_c(\Delta s \underline{1}_c)) \\ &= \mathbf{det}(\mathbb{A}_c(\underline{s})) \cdot \mathbf{det}\left(\operatorname{diag}(\underbrace{\mathbb{A}_c(\Delta s)}_{c\times 1 \text{ vector}})\right) \\ &= 0 \end{aligned} \tag{6.111}$$

and is hence singular. This implies that $\check{\mathbb{A}}$ is rank deficient and therefore the set $\underline{\check{s}}$ is also an ambiguous set. Furthermore the ranks of ambiguity of $\underline{s}$ and of $\underline{\check{s}}$ are the same. This is because the submatrices of $\mathbb{A}(s)$ with less than c rows are non-singular which implies that the submatrices of $\check{\mathbb{A}}$ with less than c rows are also non-singular. □

Chapter 7

More on Ambiguities: Symmetrical Arrays

In contrast with the type of ambiguities identified in the previous chapter which are based on partitioning of a manifold curve into *equal* segments, this chapter is concerned with the partitioning of the manifold curve into *unequal* segments from which a new type of ambiguity existing only in symmetrical[1] array structures can be identified. That is, this type of ambiguity exists in all symmetric linear and planar arrays as well as non-symmetric planar arrays which may have at least one p-curve ($p = \varphi, \alpha, \beta$) corresponding to a symmetric ELA.

7.1 Symmetric Linear Arrays and $\det(\mathbb{A}_N(\underline{s}))$

Definition 7.1 Symmetric Linear Array: A linear array of N sensors is said to be symmetric, if there is a real number b, such that the elements of the vector $\underline{\widetilde{r}} \triangleq \underline{r} + b\underline{1}_N$ satisfy the following equation

$$\text{sum}(\underline{\widetilde{r}}^n) = 0, \forall n \text{ odd} \tag{7.1}$$

Consider a set $\underline{s}$ of N arc lengths

$$\underline{s} = [s_1, s_2, ..., s_N]^T \tag{7.2}$$

on the manifold of a symmetric linear array of N sensors.

Let $\mathbb{A}_N \triangleq \mathbb{A}_N(\underline{s}) \in \mathfrak{C}^{N \times N}$ be the matrix with columns the manifold vectors corresponding to the elements of $\underline{s}$:

$$\mathbb{A}_N(\underline{s}) = [\underline{\mathbf{a}}(s_1), \underline{\mathbf{a}}(s_2), \ldots, \underline{\mathbf{a}}(s_N)] \tag{7.3}$$

[1] Every sensor has a symmetric counterpart with respect to the the array centroid.

As we have seen in Chapter 2, at every point along the manifold curve of a symmetric linear array, a set of N unity-norm coordinate vectors $\mathbb{U}(s) = [\underline{u}_1(s), \ldots, \underline{u}_N(s)] \in \mathfrak{C}^{N\times N}$ and $(N-1)$ curvatures $\kappa_1, \ldots, \kappa_{N-1}$ can be defined according to the first order matrix differential equation

$$\mathbb{U}'(s) = \mathbb{U}(s)\mathbb{C} \tag{7.4}$$

whose solution is given by

$$\mathbb{U}(s) = \mathbb{U}(0)\overbrace{\operatorname{expm}(s\mathbb{C})}^{\triangleq \mathbb{F}(s)} \tag{7.5}$$

where $\mathbb{C} \in \mathcal{R}^{N\times N}$ is the Cartan matrix, which is a real skew-symmetric matrix of the curvatures. Using Eq. (2.63) in conjunction with Eq. (7.5), the array manifold matrix $\mathbb{A}_N$ can be rewritten as

$$\begin{aligned}\mathbb{A}_N(\underline{s}) &= [\mathbb{U}(0)\mathbb{F}(s_1)\underline{R}, \mathbb{U}(0)\mathbb{F}(s_2)\underline{R}, \ldots, \mathbb{U}(0)\mathbb{F}(s_N)\underline{R}] \\ &= \mathbb{U}(0)[\mathbb{F}(s_1)\underline{R}, \mathbb{F}(s_2)\underline{R}, \ldots, \mathbb{F}(s_N)\underline{R}]\end{aligned}$$

The matrix $\mathbb{A}_N(\underline{s})$ is a square matrix and, therefore, its determinant can be used to examine the linear dependence of its columns. An expression for the determinant of $\mathbb{A}_N(s)$ can be provided as follows:

$$\det(\mathbb{A}_N(\underline{s})) = \det(\mathbb{U}(0)) \cdot \det([\mathbb{F}(s_1)\underline{R}, \mathbb{F}(s_2)\underline{R}, \ldots, \mathbb{F}(s_N)\underline{R}]) \tag{7.6}$$

Noting that for a symmetric linear array, the columns of $\mathbb{U}(s)$ are mutually orthogonal, that is

$$\mathbb{U}^H\mathbb{U} = \mathbb{I}_N \quad \text{and} \quad \det(\mathbb{U}(s)) = \pm 1 \tag{7.7}$$

and therefore the determinant of $\mathbb{A}_N$ becomes

$$\det(\mathbb{A}_N(\underline{s})) = \pm\det[\mathbb{F}(s_1)\underline{R}, \mathbb{F}(s_2)\underline{R}, \ldots, \mathbb{F}(s_N)\underline{R}] \tag{7.8}$$

Because of the special structures of the radii vector $\underline{R}$ and of the Frame matrix $\mathbb{F}$, the $\det(\mathbb{A}(\underline{s}))$ can be further written as

$$\det(\mathbb{A}_N(\underline{s})) = \pm\det\begin{bmatrix}\underline{R}^T\mathbb{F}(s_1)^T \\ \underline{R}^T\mathbb{F}(s_2)^T \\ \cdots \\ \underline{R}^T\mathbb{F}(s_N)^T\end{bmatrix}$$

$$= \pm\det\begin{bmatrix} \underline{R}^T\mathbb{F}(-s_1) \\ \underline{R}^T\mathbb{F}(-s_2) \\ \dots \\ \underline{R}^T\mathbb{F}(-s_N) \end{bmatrix}$$

$$= \pm\det\begin{bmatrix} \underline{R}^T\mathbb{F}(s_1) \\ \underline{R}^T\mathbb{F}(s_2) \\ \dots \\ \underline{R}^T\mathbb{F}(s_N) \end{bmatrix} \tag{7.9}$$

From Eq. (2.61) evaluated at $s = 0$ and for a symmetric linear array:

$$R^T = a^H(0)\mathbb{U}(0) \tag{7.10}$$

and therefore,

$$\det(\mathbb{A}_N(\underline{s})) = \pm\det\begin{bmatrix} \underline{a}(0)^H\mathbb{U}(0)\mathbb{F}(s_1) \\ \underline{a}(0)^H\mathbb{U}(0)\mathbb{F}(s_2) \\ \vdots \\ \underline{a}(0)^H\mathbb{U}(0)\mathbb{F}(s_N) \end{bmatrix}$$

$$= \pm\det\begin{bmatrix} \underline{a}(0)^H\mathbb{U}(s_1) \\ \underline{a}(0)^H\mathbb{U}(s_2) \\ \vdots \\ \underline{a}(0)^H\mathbb{U}(s_N) \end{bmatrix} \quad \text{(using Eq. (7.5))}$$

$$= \pm\det\begin{bmatrix} \underline{a}(0)^H\underline{u}_1(s_1) & \underline{a}(0)^H\underline{u}_2(s_1) & \dots & \underline{a}(0)^H\underline{u}_N(s_1) \\ \underline{a}(0)^H\underline{u}_1(s_2) & \underline{a}(0)^H\underline{u}_2(s_2) & \dots & \underline{a}(0)^H\underline{u}_N(s_2) \\ \vdots & \vdots & \vdots & \vdots \\ \underline{a}(0)^H\underline{u}_1(s_N) & \underline{a}(0)^H\underline{u}_2(s_N) & \dots & \underline{a}(0)^H\underline{u}_N(s_N) \end{bmatrix} \tag{7.11}$$

7.2 Characteristic Points on the Array Manifold

An initial partition of a hyperhelical curve into $N-1$ (or more) unequal segments can be achieved using the concept of characteristic points, defined as follows:

Definition 7.2 Characteristic Points: If $\mathcal{A}$ is the manifold of an array of N sensors with locations $\underline{r}$, a point s on $\mathcal{A}$, with

$$0 \leq s \leq l_{\mathrm{m}} = 2\pi\|\underline{r}\| \tag{7.12}$$

is a characteristic point if and only if

$$\mathrm{Re}\{\underline{\mathbf{a}}^H(0)\underline{u}_1(s)\} = 0 \tag{7.13}$$

where $\underline{u}_1(s)$ denotes the tangent vector at the manifold point s, i.e. $\underline{u}_1(s) = \underline{\mathbf{a}}'(s)$.

For a general linear array the above definition indicates an important property of characteristic points. The tangent at any characteristic point is orthogonal (wide sense orthogonality since *only* the real part is zero) to the manifold vector at $s = 0$. This orthogonality, in the case of a *symmetric* linear, becomes "narrow" sense. Thus, if the manifold of a *symmetric* linear array has N (or more) characteristic points $s_{c_1}, s_{c_2}, \ldots, s_{cN}$, where $s_{c_1} = 0$, then this (denoted by $\underline{s}_{cp}$) set of characteristic points is *ambiguous*, as the first column of Eq. (7.11) is the all zero vector, implying that

$$\mathrm{rank}(\mathbb{A}_N(\underline{s}_{cp})) = N - 1 \quad \text{and} \quad \det(\mathbb{A}_N(\underline{s}_{cp})) = 0 \tag{7.14}$$

This is the starting point for the identification of a new class of ambiguous generator set having a rank of ambiguity ρ_{a} equal to $(N - 1)$.

The existence, however, of at least N characteristic points within the manifold length l_{m} is very important for the identification of ambiguous situations and this, in turn, relates to the number of windings. In particular we have seen in Chapter 2 (Theorem 2.2, page 41) that the manifold of a linear array of N sensor has a hyperhelical shape with the number of windings ($N =$ odd) or half-windings ($N =$ even) n_w given by the following equation

$$n_w = \frac{l_{\mathrm{m}}}{l_w} \tag{7.15}$$

where l_w is the length of one winding or one half-winding and is given by the $(N - 1)$th positive root of the following expression:

$$\mathrm{Tr}(\mathbb{C}\ \mathrm{expm}(s\mathbb{C})) = 0 \tag{7.16}$$

This expression and its derivatives with respect to s are the pivots for identifying ambiguities which are only present in symmetric linear arrays (or symmetric ELAs).

Equation (7.15) indicates that when $n_w > 1$, then there is more than one winding (or half-winding) contained in the manifold length of the curve. It must be noted that uniform linear arrays with intersensor spacing equal to one half-wavelength have $l_{\mathrm{m}} = l_w \Longrightarrow n_w = 1$. A spiral of a spring could be used to provide a visualization of the concept of a winding.

Corollary 7.1 *A* ***sufficient*** *condition for the presence of nontrivial ambiguities in the manifold $\mathcal{A}$ of a* **symmetric** *linear array (or ELA) is $l_{\mathrm{m}} > l_w$ i.e. $n_w > 1$*

The above implies that if $0° \le p_1 < p_N \le 180°$ and

$$n_w = \frac{l_w}{2\pi\|\underline{r}\|} \geqslant \frac{2}{|\cos p_1 - \cos p_N|} \tag{7.17}$$

then there exist $N-2$ bearings $p_2, \ldots, p_{N-1}$ in the region (p_1, p_N) such that the manifold vectors corresponding to $p_1, p_2, \ldots, p_{N-1}$ are *linearly dependent.*

Example 7.1 Consider a symmetric linear array of $N = 4$ sensors with locations given by the following vector in half-wavelengths:

$$\underline{r} = [-1.9, -0.6, 0.6, 1.9]^T \tag{7.18}$$

The length of the hyperhelical manifold curve of this array is $l_{\mathrm{m}} = 2\pi\|\underline{r}\| = 17.7048$ and its Cartan matrix

$$\mathbb{C} = \begin{bmatrix} 0 & -0.6462 & 0 & 0 \\ 0.6462 & 0 & -0.1819 & 0 \\ 0 & 0.1819 & 0 & -0.2222 \\ 0 & 0 & 0.2222 & 0 \end{bmatrix} \tag{7.19}$$

Figure 7.1 is the plot of Eq. (7.16) with respect to the arc length. From this plot it can be seen that the manifold curve consists of more than one half-winding. Actually, the length of one half-winding is equal to $l_w = 14.0476$, i.e. the arc length at the $(N-1)$th positive characteristic point. The number of half-windings of the hyperhelical curve is also calculated using Eq. (7.15) and is equal to $n_w = 1.2603$. The first four roots (arc lengths) of Eq. (7.16) are

$$\underline{s} = [0, 5.0780, 8.8664, 14.0476]^T \tag{7.20}$$

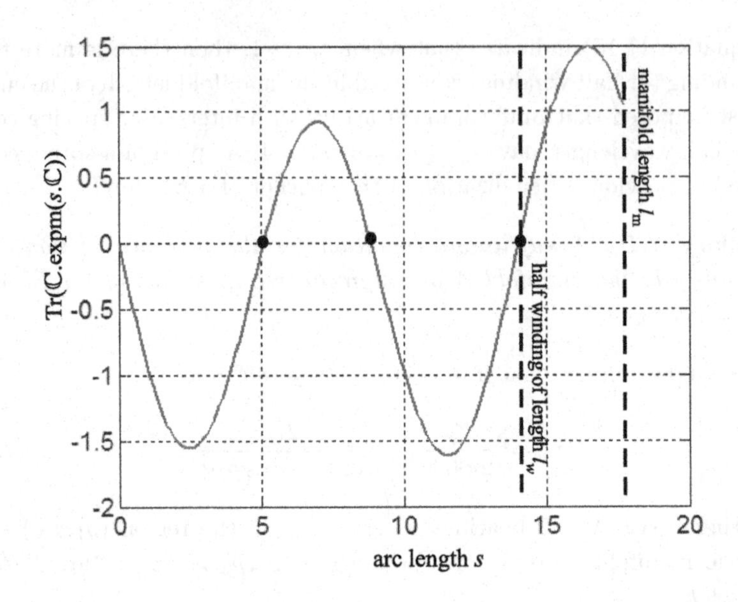

Fig. 7.1 Plot of $\mathrm{Tr}(\mathbb{C}\,\mathrm{expm}(s\mathbb{C}))$ with respect to s for the array of Eq. (7.18).

corresponding to the following set of azimuth angles (i.e. the directions of arrival)

$$\underline{\theta} = [0°, 64.7614°, 90.0919°, 125.9361°]^T \tag{7.21}$$

Note that the above set $\underline{s}$ is an ambiguous set. i.e.

$$\mathrm{rank}(\mathbb{A}(\underline{s})) = N - 1 \quad \text{and} \quad \det(\mathbb{A}(\underline{s})) = 0 \tag{7.22}$$

Furthermore the set $\underline{s}$ has its first element equal to zero and also satisfies all the other conditions of the definition of the ambiguous generator set (see Definition 6.5, page 120). Hence $\underline{s}$ is also an ambiguous generator set of parameters.

It is clear from the above discussion that if the array is symmetric, in addition to the AGSs constructed from the Uniform Basic Sets of Eq. (6.23), more AGSs can be found using a non-uniform partition of the array manifold based on the roots of Eq. (7.16). However, the roots of this equation are not the only ones producing AGSs. In fact we will see in the next section that the derivatives of the above equation produce an infinite number of

AGSs and this complicates the representation and identification of this type of ambiguity.

7.3 Array Symmetricity and Non-Uniform Partitions of Hyperhelices

Before we proceed let us generalize the concept of characteristic points to "characteristic points of order n" by rewriting the inner product $\underline{\mathbf{a}}^H(0)\underline{u}_1(s)$ as follows:

$$\begin{aligned}\underline{a}(0)^H\underline{u}_1(s) &= \frac{d}{ds}(\underline{a}(0)^H\underline{a}(s)) \\ &= \frac{d}{ds}(\underline{R}^T\mathbb{U}(0)^H\mathbb{U}(0)\mathbb{F}(s)\underline{R}) \\ &= \frac{d}{ds}\mathrm{Tr}\left(\underbrace{\mathbb{U}(0)^H\mathbb{U}(0)}_{=\mathbb{I}_N}\mathbb{F}(s)\underline{R}\,\underline{R}^T\right) \\ &= \frac{d}{ds}\mathrm{Tr}(\mathbb{F}(s)\underline{R}\,\underline{R}^T) \\ &= \frac{d}{ds}\mathrm{Tr}(\mathbb{F}(s)) \\ &= \mathrm{Tr}(\mathbb{F}'(s))\end{aligned}$$

i.e.

$$\underline{a}(0)^H\underline{u}_1(s) = \mathrm{Tr}(\mathbb{C}\,\mathrm{expm}(s\mathbb{C})) \tag{7.23}$$

where the expression $\mathbb{F}(s) = \mathbb{C}\mathrm{expm}(s\mathbb{C})$ has been used. Then by differentiating Eq. (7.23) we get

$$\underline{a}(0)^H\underline{u}_1^{\prime(n-1)}(s) = \mathrm{Tr}(\mathbb{C}^n\,\mathrm{expm}(s\mathbb{C})), \quad n = 1, 2, \ldots \tag{7.24}$$

where $\underline{u}_1^{\prime(n)}$ denotes the nth derivative of the tangent vector at the manifold point s. This leads to the following definition:

Definition 7.3 Consider a symmetric linear array of N sensors having a manifold $\mathcal{A}$ of length l_m and a Cartan matrix $\mathbb{C}$. A point s on $\mathcal{A}$ with

$$0 \leq s \leq l_\mathrm{m} = 2\pi\|\underline{r}\| \tag{7.25}$$

is said to be a characteristic point of order n if and only if

$$\mathrm{Tr}(\mathbb{C}^n\,\mathrm{expm}(s\mathbb{C})) = 0 \tag{7.26}$$

The characteristic points of order 1 ($n = 1$), i.e.

$$\underline{a}(0)^H \underline{u}_1(s) = \text{Tr}(\mathbb{C}\ \text{expm}(s\mathbb{C})) = 0 \tag{7.27}$$

will simply be referred to as "characteristic points."

Theorem 7.1 ***Non Uniform Partition Type Ambiguities:*** *Consider a symmetric linear array (or an ELA) of N sensors. If there are $c \geqslant N$ characteristic points of order n on the array manifold $\mathcal{A}$, forming the vector*

$$\underline{s}_{cp}^{(n)} = [s_1, s_2, \ldots, s_c]^T \tag{7.28}$$

that is, the elements of $\underline{s}_{cp}^{(n)}$ are the first n roots of the function

$$\text{Tr}(\mathbb{C}^n\ \text{expm}(s\mathbb{C})) = 0 \tag{7.29}$$

then $\underline{s}_{cp}^{(n)}$, as well as any of its subsets with N elements, constitute a rank $\rho_a = (N-1)$ ambiguous set of arc lengths.

Theorem 7.1 can be easily proven for $n \leqslant N$, using the recursive equation Eq. (2.12) while for $n > N$, $n \in \mathcal{Z}^+$ it can be proven using the Cayley–Hamilton Theorem, which states that "every matrix satisfies its own characteristic equation," as applied to the Cartan matrix.

However not all the existing ambiguous generator sets are calculated by using the set of roots, between zero and the manifold length. Roots that are bigger than the manifold length should also be taken into account due to the non uniform segmentation of the manifold. Furthermore the roots of Eq. (7.29) are ambiguous for **any real value** of n greater than 0, as shown by the following example

Example 7.2 For an array with sensor positions

$$\underline{r} = [-2.1, -0.9, 0.9, 2.1]^T \tag{7.30}$$

the following table gives the AGSs produced from the Uniform Basic Sets

s_1	s_2	s_3	s_4	ρ_a
0	6.7672	13.5344	—	2
0	4.8337	9.6674	14.5011	3
0	4.8337	9.6674	19.3349	3
0	4.8337	14.5011	19.3349	3
0	9.6674	14.5011	19.3349	3

Furthermore the Cartan Matrix $\mathbb{C}$ of this array is

$$\mathbb{C} = \begin{bmatrix} 0 & -0.6074 & 0 & 0 \\ 0.6074 & 0 & -0.2056 & 0 \\ 0 & 0.2056 & 0 & -0.2981 \\ 0 & 0 & 0.2981 & 0 \end{bmatrix} \tag{7.31}$$

and since it is a symmetric array, the roots of Eq. (7.29), (for $n = 1$) i.e. the set of characteristic points,

$$\underline{s}_{cp}^{(1)} = [0, 5.5147, 9.3228, 14.0360, 19.8000]^T$$

provide the four "non-uniform" AGS of rank $\rho_{\mathrm{a}} = 3$ as shown below:

s_1	s_2	s_3	s_4	ρ_{a}
0	5.5147	9.3229	14.0360	3
0	5.5147	9.3229	19.8000	3
0	5.5147	14.0360	19.8000	3
0	9.3229	14.0360	19.8000	3

However for $n = 1.9$ the first four roots of Eq. (7.29) for $s \in [0, l_{\mathrm{m}}]$ forming the vector

$$\underline{s}_{cp}^{(1.9)} = [2.9041, 7.6098, 12.0167, 17.1851]^T \tag{7.32}$$

is also an ambiguous set and of rank $N - 1 = 3$. Eq. (7.29) has been plotted in Fig. 7.2 for $n = 1.9$.

From Example 7.2 it can be observed that the first root of Eq. (7.29) is not equal to zero for all values of n. This means that although in the case of $n = 1$ the AGSs can be simply identified as all the combinations of three, or in general $N - 1$, non-zero elements of $\underline{s}_{cp}^{(1)}$ with a zero prepended, the same cannot be done for $n = 1.9$. In order to find the AGSs in this case, the first element of the ambiguous set must be subtracted from all its elements.

Note that the first root of Eq. (7.29) is equal to zero only if $n = 1, 3, 5$, etc. This can be seen from Eq. (7.29) which for $s = 0$ becomes equal to $\mathrm{Tr}(\mathbb{C}^n)$ that is zero only if n is odd, since in that case its structure is identical with the structure of $\mathbb{C}$ (zero diagonal, more precisely only its two secondary diagonals are non-zero). This result is confirmed by Fig. 7.3

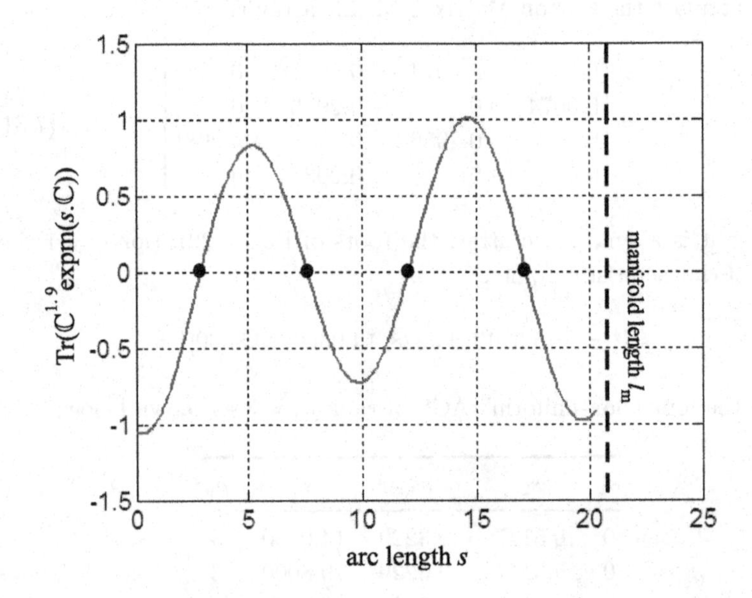

Fig. 7.2 Plot of $\mathrm{Tr}(\mathbb{C}^n \operatorname{expm}(s\mathbb{C}))$ for $n = 1.9$.

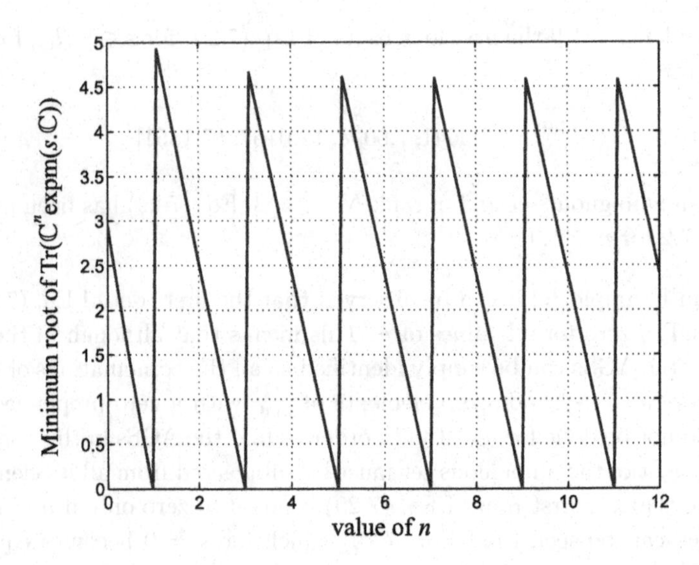

Fig. 7.3 Plot of smallest root of $\mathrm{Tr}(\mathbb{C}^n \operatorname{expm}(s\mathbb{C}))$ versus n for the array in Example 7.2.

where the first root of $\text{Tr}(\mathbb{C}^n \text{expm}(s\mathbb{C}))$ is plotted for $n = 0.1$ to 10 in steps of 0.1 for the Cartan matrix of the array in Example 7.2.

From the previous discussion it is clear that an infinite number of AGSs exist in symmetric linear arrays because n takes infinite values. However for large values of n there exists a relationship between Theorems 6.2 and 7.1. This is stated in the following lemma:

Lemma 7.1 *For $n \to \infty$ (large), the roots of $\text{Tr}(\mathbb{C}^n \text{expm}(s\mathbb{C})) = 0$, $n > 0$ are a shifted version of the set of arc lengths created by partitioning the manifold in segments equal to l_m/l_a, where $l_a = \Delta r_{1,N}$ is the aperture of the array.*

Example 7.3 Consider the array

$$\underline{r} = [2.6870, 1.2728, 0, -1.2728, -2.6870]^T \tag{7.33}$$

Using the aperture intersensor spacing $\Delta r_{1,5} = 5.3740$ the UBS $\underline{s}_{\Delta r_{1,5}}$ can be found as

$$\underline{s}_{\Delta r_{1,5}} = [0, 4.9162, 9.8322, 14.7483, 19.6645, 24.5807]^T \tag{7.34}$$

The set of characteristic points $\underline{s}_{cp}^{(n)}$ of order $n > 6$, given by using Eq. (7.29) of Theorem 7.1, has approximately the same elements as $\underline{s}_{\Delta r_{1,5}}$. For instance for $n = 13$

$$\underline{s}_{cp}^{(13)} = [0, 4.9162, 9.8322, 14.7483, 19.6645, 24.5807]^T \tag{7.35}$$

indicating that the two sets are almost identical, i.e.

$$\underline{s}_{cp}^{(13)} \simeq \underline{s}_{\Delta r_{1,5}} \tag{7.36}$$

7.4 Ambiguities of Rank-$(N-1)$ and Array Pattern

In the previous section we have seen that the function $\text{Tr}(\mathbb{C}\text{expm}(s\mathbb{C}))$ is central for the identification and generation of non-uniform rank-$(N-1)$ ambiguous sets of arc lengths in symmetric linear arrays. This can be also rewritten as:

$$\text{Tr}(\mathbb{C}\,\text{expm}(s\mathbb{C})) \equiv \underline{\mathbf{a}}(0)^H \underline{u}_1(s) = \frac{d}{ds}\left(\underline{\mathbf{a}}(0)^H \underline{\mathbf{a}}(s)\right) \tag{7.37}$$

where the term $\underline{\mathbf{a}}(0)^H \underline{\mathbf{a}}(s)$ is purely imaginary and represents the *array pattern* of an array steered towards endfire, as a function of the arc length s.

It is immediately apparent that the roots of $\text{Tr}(\mathbb{C}\,\text{expm}(s\mathbb{C})) = 0$ are equal to the arc lengths corresponding to the stationary points of the gain pattern steered towards endfire and this leads to the following proposition.

Proposition 7.1 *If the array pattern of an N-sensor symmetric linear array, parametrized in terms of arc length, steered towards endfire exhibits more than N stationary points (usually lobes), then the corresponding directions are rank-$(N-1)$ ambiguous.*

The following example illustrates the main characteristics of the above proposition:

Example 7.4 Consider a symmetric linear array with sensor locations in half-wavelengths given by

$$\underline{r} = [-1.45, -0.8, 0.8, 1.45]^T \tag{7.38}$$

and with manifold length $l_m = 14.7153$. The function $\text{Tr}(\mathbb{C}\,\text{expm}(s\mathbb{C}))$ is plotted in Fig. 7.4 for $s \in [0, l_m]$, which is equivalent to the azimuth space $\theta \in [0°, 180°)$. As can be seen, the equation $\text{Tr}(\mathbb{C}\,\text{expm}(s\mathbb{C})) = 0$ has 4 roots within the manifold length given by

$$\underline{s}_{cp}^{(1)} = [0,\ 5.9142,\ 10.5499,\ 14.2941] \tag{7.39}$$

which constitutes a rank-3 AGS. The corresponding set of rank-3 ambiguous set of directions is given by

$$\underline{\theta}_{cp}^{(1)} = [0°,\ 78.69°,\ 115.71°,\ 160.52°] \tag{7.40}$$

Furthermore, as can be seen from Fig. 7.5, the gain pattern of the array steered towards endfire exhibits stationary points at

$$\underline{s} = [0,\ 5.9142,\ 10.5499,\ 14.2941] \tag{7.41}$$

when parametrized in terms of the arc length, and at

$$\underline{\theta} = [0°,\ 78.69°,\ 115.71°,\ 160.52°] \tag{7.42}$$

when parametrized with respect to the azimuth. It is easy to verify that s and θ are equivalent. It might however be argued that a source close to endfire is unlikely; but it must be recalled that any displacement of the ambiguous set of arc lengths by an arbitrary amount will result in another ambiguous set, provided that the manifold length is not exceeded.

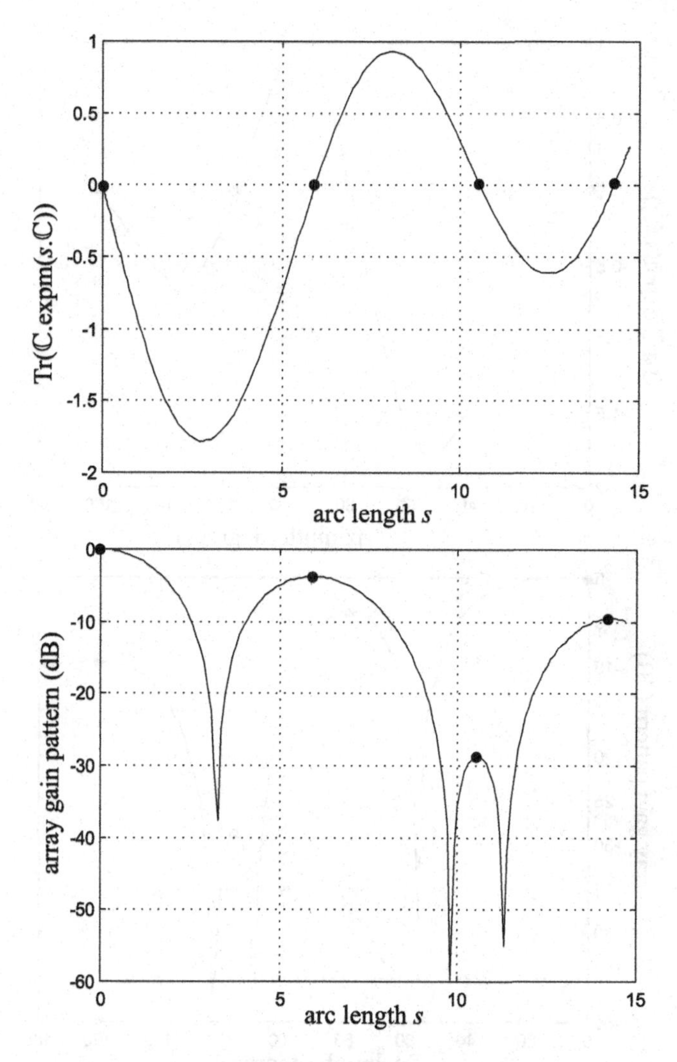

Fig. 7.4 The function Tr($\mathbb{C}$expm($s\mathbb{C}$)) and the array pattern, plotted as a function of the arc length, for a symmetric array with locations $\underline{r} = [-1.45, -0.8, 0.8, 1.45]^T$. Note that the arc length space $s \in [0, l_{\mathrm{m}} = 14.7153)$ is equivalent to the azimuth space $\theta \in [0°, 180°)$, while the black dots constitute a rank-3 AGS of arc lengths.

For instance a displacement of the set $\underline{s}$ by $\Delta s = 0.3$, corresponding to a new set of directions given by

$$\theta = [16.42°, 81.06°, 118.34°, 169.59°] \tag{7.43}$$

is also ambiguous.

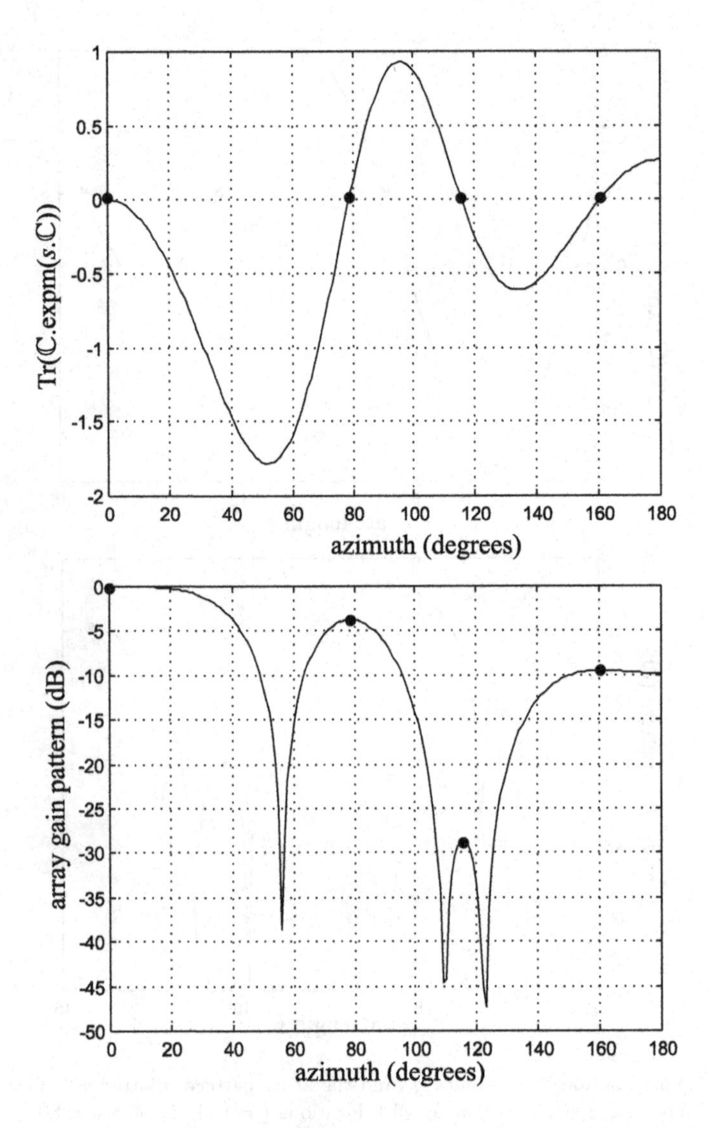

Fig. 7.5 The results of Fig. 7.4 expressed as a function of azimuth.

7.5 Planar Arrays and 'Non-Uniform' Ambiguities

The technique presented in the previous section to identify ambiguities arising from non-uniform partitions of the symmetric linear array manifold can be readily applied to any hyperhelical curve (e.g. a composite ϕ-curve) of the planar array manifold if the corresponding equivalent linear array (ELA) is symmetric.

Note that if the planar array itself is symmetric, then all ELAs of the family of p-curves are symmetric too and this is formally stated as follows:

Corollary 7.2 *All equivalent linear arrays $\underline{r}(\Psi_0), \forall\Psi_0$ (see Eq. 6.4) associated with the families of p-curves ($p = \varphi, \alpha, \beta$) of a symmetric planar array are symmetric and therefore can suffer from both uniform and non-uniform types of ambiguity.*

Note that ambiguities resulting from non-uniform partitions of the array manifold are present only in symmetric linear arrays and hence this class of ambiguities is applicable only when an ELA is symmetric. However, the prerequisite symmetry of the ELA does not impose any special restriction on the planar array configuration, and an arbitrary planar array may exhibit symmetric ELAs along certain Ψ_o angles. However, if the planar array is symmetric about both the x- and y-axes or any two orthogonal directions, then the ELA along any Ψ_o angle is symmetric. Hence symmetric planar arrays, like symmetric linear arrays, are more prone to ambiguities. If the planar array is asymmetric, then it is more likely to suffer from ambiguities along Ψ_o angles where the ELA is symmetric than those where it is not.

Consider the issue of collocation of projected sensors in a symmetric ELA. For instance an N-sensor ELA consisting of an underlying symmetric $(N - \ell)$-sensor linear array and ℓ sensors collocated with other sensors. If the underlying $(N-\ell)$-sensor array suffers from rank-$(N-\ell-1)$ ambiguity resulting from non-uniform partitions of the array manifold, then it is easy to see that the N-sensor ELA will also suffer from the same rank-$(N-\ell-1)$ ambiguity, with the same ambiguous directions. In fact, the N-sensor ELA will suffer from any ambiguity inherent to the underlying $(N - \ell)$-sensor linear array. For example, consider a planar array whose sensor locations in half-wavelengths are given by

$$\mathbf{r} = \begin{bmatrix} -0.9548 & -0.6623 & 0.1077 & 0.5174 & 0.6168 \\ -1.2453 & 1.0057 & -1.5300 & -0.9669 & 1.3359 \\ 0 & 0 & 0 & 0 & 0 \end{bmatrix}^T$$

The ELA along $\theta_o = 75°$ is given by

$$\underline{r}(\theta_o) = [-1.45, -1.45, -0.8, 0.8, 1.45]^T \tag{7.44}$$

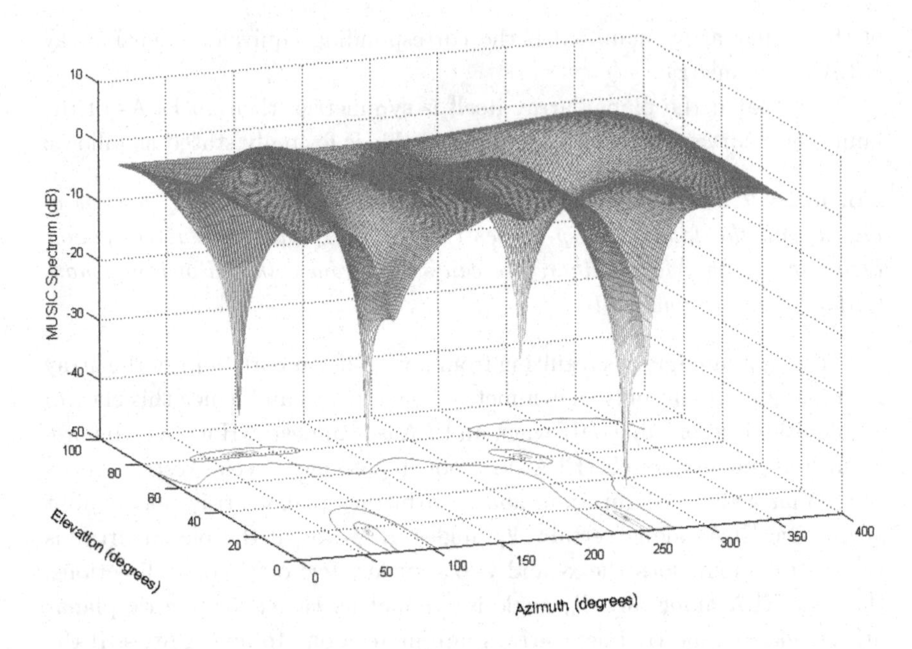

Fig. 7.6 Two-dimensional MUSIC in an ambiguous signal environment.

with two collocated sensors at -1.45. Discarding one of the collocated sensors, the resulting array is found to be symmetric and given by

$$\underline{r} = [-1.45, -0.8, 0.8, 1.45]^T \tag{7.45}$$

Note that this is precisely the array examined in Example 7.4 and hence the following set of directions will constitute an ambiguous set of directions for the corresponding planar array:

$$(\underline{\theta}, \underline{\phi})_{cp}^{(1)} = \{(75°, 16.42°), (75°, 81.06°), (255°, 61.66°), (255°, 10.41°)\}$$

Figure 7.6 illustrates the contour plot of the MUSIC spectrum generated using 100 data snapshots collected from the planar array, operating in the presence of three equipower uncorrelated sources with SNR $= 10\,\text{dB}$ and impinging from directions $(75°, 16°), (75°, 81°)$ and $(255°, 62°)$ respectively. A spurious peak around the $(255°, 10°)$ direction is also observed.

7.6 Conclusions

Although all "uniform" ambiguities can be found using the Uniform Basic Sets of arc lengths of Eq. (6.23), there is no single coherent framework at this moment to identify and estimate *all* "non-uniform" ambiguities. However, this chapter provides a starting point by presenting a class of rank-$(N-1)$ non-uniform ambiguities. Since the range of n is infinite it can hence be concluded that an infinite number of AGSs exist in symmetric linear and symmetric planar arrays. This is also true for non-symmetric planar arrays if at least one of its ELAs is "symmetric." Note that each AGS represents an infinite number of ambiguous sets — rotated in arc length (see Theorem 6.1). Having an infinite rather than a finite number of AGSs is currently a fundamental limitation of the theory for handling ambiguities in symmetric arrays. Thus further research effort should be devoted to the generalization of the concept of AGSs in order to reduce their number to finite values. Furthermore the exploitation of any inherent relationship between the various AGSs, constructed from the characteristic points of different order n, may play a crucial role in dictating the identification, classification and estimation of this huge number of ambiguities inherent in symmetric array structures.

Chapter 8

Array Bounds

It was explained in Chapter 2 that, according to standard differential geometry, a d-dimensional curve may be fully described via $d-1$ curvatures. However, it is intuitively apparent that a sufficiently small segment of a multi-dimensional curve $\mathcal{A}$ may in fact be accurately approximated by a *circular arc*. Despite the fact that visualization of a curve of dimensionality greater than three is impossible, **circular approximation provides an informative notion of the curve's shape in the local neighborhood of an arbitrary point** $\breve{s}$ and is defined in terms of the curve's principal curvature only, thereby simplifying the analysis considerably. In this chapter the circular approximation of a curve $\mathcal{A}$ is used to determine the array's ultimate accuracy, detection and resolution capabilities. Note that the curve $\mathcal{A}$ could be the manifold of a linear array or a p-curve ($\mathcal{A}_{p|q_o}$ with $p = \theta, \phi, \alpha$ or β), lying on the manifold surface $\mathcal{M}$ of a planar array and associated with an ELA.

8.1 Circular Approximation of an Array Manifold

A sufficiently small neighborhood of a point $\breve{s}$ on a curve in real N-dimensional space (where the coordinate vectors are strictly orthogonal) can be approximated by a circular arc. This circular arc lies on

$$\mathcal{H}_{12} \triangleq \mathcal{L}([\underline{u}_1(\breve{s}), \underline{u}_2(\breve{s})]) \quad \text{(i.e. a plane)} \tag{8.1}$$

with the vector $\underline{u}_1(\breve{s})$ as its tangent, and has radius R which is equal to the inverse of the curvature of the curve at $\breve{s}$. This implies that

$$\underline{u}_1(\breve{s} \pm \Delta s_i) = \underbrace{\cos\left(\frac{\Delta s_i}{R}\right)}_{\simeq\left(1-\frac{\Delta s_i^2}{2R^2}\right)} \underline{u}_1(\breve{s}) \pm \underbrace{\sin\left(\frac{\Delta s_i}{R}\right)}_{\simeq\frac{\Delta s_i}{R}} \underline{u}_2(\breve{s}) \tag{8.2}$$

Here the above concept is applied in order to derive a "circular approximation" to the manifold curve $\mathcal{A}$ which is embedded in an N-dimensional complex space and is defined as follows:

$$\mathcal{A} = \{\underline{\mathbf{a}}(p) \in \mathfrak{C}^N, \ \forall p : p \in \Omega_p\} \tag{8.3}$$

Let us consider an arbitrary point $\breve{s}$ (arc length) on the manifold and form the Frame matrix $\mathbb{F}$ at $\breve{s}-\Delta s_1$ and $\breve{s}+\Delta s_2$ corresponding to parameter values $\breve{p}-\Delta p_1$ and $\breve{p}+\Delta p_2$. Since $\mathbb{F}(\breve{s}) = \text{expm}(\breve{s}\mathbb{C})$, by using Taylor series expansion and retaining terms up to second order:

$$\mathbb{F}(\breve{s} \pm \Delta s_i) \simeq \mathbb{F}(\breve{s})\left(\mathbb{I}_d \pm (\Delta s_i)\mathbb{C} + \frac{1}{2}(\Delta s_i)^2\mathbb{C}^2\right) \quad \text{for } i = 1, 2 \tag{8.4}$$

However the matrix of the manifold coordinate vectors $\mathbb{U}$ at $\breve{s} \pm \Delta s_i$ can be expressed (see Eq. (2.10), page 28) as

$$\mathbb{U}(\breve{s} \pm \Delta s_i) = \mathbb{U}(0)\mathbb{F}(\breve{s} \pm \Delta s_i) \tag{8.5}$$

Combining Eqs. (8.4) and (8.5), and using the special skew symmetric form of the Cartan matrix (Eq. (2.14)), the following approximation to the manifold tangent vector at $\breve{s} \pm \Delta s_i$ (i.e. the first column of $\mathbb{U}(\breve{s} \pm \Delta s_i)$) may be obtained:

$$\underline{u}_1(\breve{s} \pm \Delta s_i) \simeq \left(1 - \frac{(\Delta s_i)^2}{2}\kappa_1^2\right)\underline{u}_1(\breve{s}) \pm \Delta s_i \kappa_1 \underline{u}_2(\breve{s}) + \frac{(\Delta s_i)^2}{2}\kappa_1\kappa_2\underline{u}_3(\breve{s}) \tag{8.6}$$

However, the 3rd term of Eq. (8.6) can be ignored so long as a tolerance factor $TOL \ll 1$ is defined such that:

$$\left|\frac{\text{3rd term}}{\text{2nd term}}\right| = \frac{1}{2}\Delta s_i \kappa_2 \leq TOL \tag{8.7}$$

where κ_2 can be found using Eq. (2.37), i.e.

$$\kappa_2 = \frac{1}{\kappa_1}\|\underline{\widetilde{r}}^3 - \kappa_1^2 \underline{\widetilde{r}}\| \tag{8.8}$$

Thus, Eq. (8.7) provides the following condition

$$\Delta s_i \leq \frac{2\kappa_1}{\|\underline{\widetilde{r}}^3 - \kappa_1^2 \underline{\widetilde{r}}\|} TOL \tag{8.9}$$

under which Eq. (8.6) is simplified to

$$\underline{u}_1(\check{s} \pm \Delta s_i) \simeq \left(1 - \frac{(\Delta s_i)^2}{2}\kappa_1^2\right) \underline{u}_1(\check{s}) \pm \Delta s_i \kappa_1 \underline{u}_2(\check{s}) \tag{8.10}$$

matching Eq. (8.2) for $R = \kappa_1^{-1}$.

The above indicates that a circular arc can be adopted as an excellent approximation to the array manifold in the region of interest neighboring $\check{s}$. This approximation also provides a 2D representation where, at a local level, a curve has its main components along the first two coordinate vectors at point $\check{s}$, as illustrated in Fig. 8.1, and has not yet had the opportunity to "move into" higher dimensions (subspace $\mathcal{L}\{[\underline{u}_3, \underline{u}_4, \cdots, \underline{u}_d]\}$).

Note that by writing the incremental arc length as $\Delta s_i \approx \dot{s}(\check{p})\Delta p_i$ and recalling that $\dot{s}(p) = \pi\|\underline{r}\| \sin p$, the condition of Eq. (8.9) provides an upper limit to the angular separation Δp_i for which the above approximation is

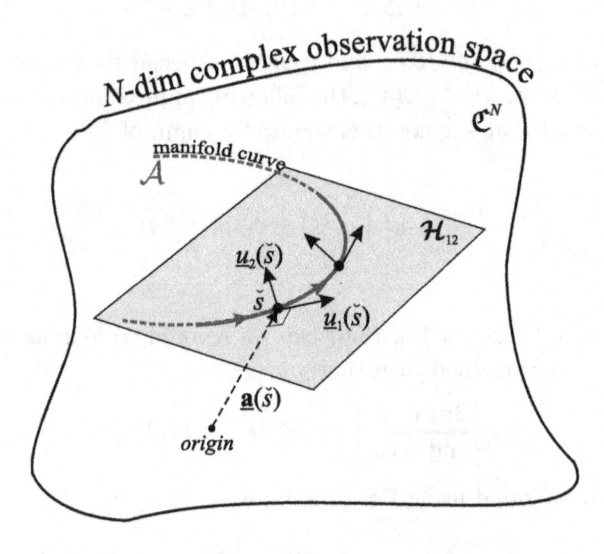

Fig. 8.1 $\mathcal{H}_{12}$ subspace used to approximate the array manifold at a local level.

valid. That is, for $i = 1, 2$ we have

$$\Delta p_i \leq \underbrace{\frac{2\kappa_1}{\pi \left\|\underline{r}\right\| \sin \breve{p} \|\underline{\widetilde{r}}^3 - \kappa_1^2 \underline{\widetilde{r}}\|}}_{=\dot{s}(\breve{p})} TOL \tag{8.11}$$

It should be emphasized that although the array manifold is a vector continuum with a constant norm $\sqrt{N}$, the radius of the circular approximation is different from $\sqrt{N}$.

Consider the Fig. 8.2 which illustrates a manifold vector $\underline{\mathbf{a}}(\breve{s} + \Delta s_i)$ in the local neighborhood of $\breve{s}$ with Δs_i equal to $-\Delta s_1$ or $+\Delta s_2$ corresponding to two signals arriving from bearings p_1 and p_2. If $\underline{u}_1(\breve{s}) \perp \underline{u}_2(\breve{s})$ then the radius of this circular arc is equal to $\kappa_1^{-1}(\breve{s})$. However, $\underline{u}_1(\breve{s})$ and $\underline{u}_2(\breve{s})$ are not, in general, strictly orthogonal. Therefore, the radius of the corresponding circular arc at a point $\breve{s}$ is equal to $\hat{\kappa}_1^{-1}(\breve{s})$ which takes into account the "orientation" of the array manifold curve. The parameter $\hat{\kappa}_1(\breve{s})$ is the curvature (inverse of radius) of the circular approximation of the manifold $\mathcal{A}$ and in Appendix 8.9.1 is proved to be

$$\hat{\kappa}_1 = \kappa_1 \sin(\zeta) \tag{8.12}$$

where ζ is the angle between $\underline{u}_1(\breve{s})$ and $\underline{u}_2(\breve{s})$. Indeed

$$\sin(\zeta) = \sqrt{1 - |\underline{u}_1^H(s)\underline{u}_2(s)|^2} \tag{8.13}$$

$$= \sqrt{1 - \left|\frac{j}{\kappa_1}\underline{1}_N^T(\underline{\widetilde{r}}^3)\right|^2} \text{ (using Eq. (2.44))} \tag{8.14}$$

Thus Eq. (8.12) becomes

$$\boxed{\begin{gathered}\hat{\kappa}_1 = \sqrt{\kappa_1^2 - \left|\underline{1}_N^T\underline{\widetilde{r}}^3\right|^2} \\ \text{where } \underline{\widetilde{r}} = \underline{r}/\left\|\underline{r}\right\|\end{gathered}} \tag{8.15}$$

Since the manifold can be approximated by a circular arc, it is clear that

$$\Delta s_i = \hat{\kappa}_1^{-1} \Delta\psi_i \tag{8.16}$$

where $\Delta\psi_i$ (with reference to Fig. 8.2) is estimated as follows:

$$\cos(\Delta\psi_i) = \frac{\hat{\kappa}_1^{-1} - \mu_{x_i}}{\hat{\kappa}_1^{-1}} \tag{8.17}$$

$$\Rightarrow \Delta\psi_i = \arccos(1 - \hat{\kappa}_1\mu_{x_i}), \quad \text{for } i = 1, 2 \tag{8.18}$$

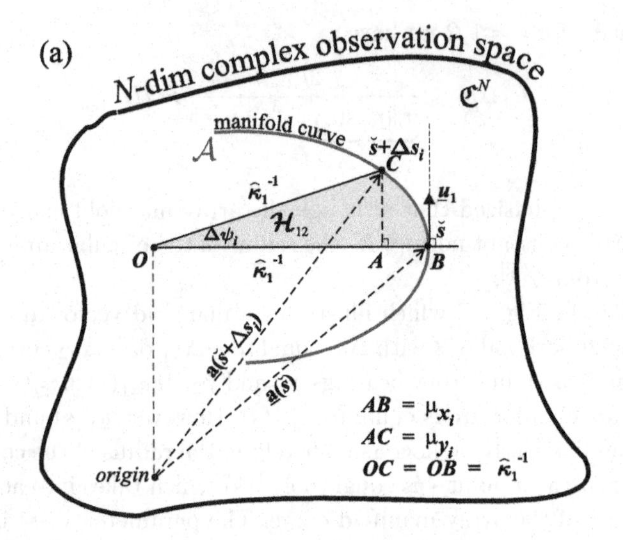

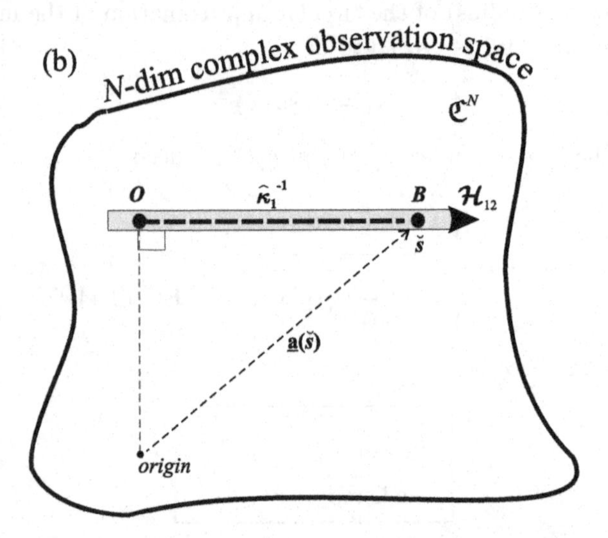

Fig. 8.2 Two alternative ways of visualizing the 'circular approximation' of array manifold curve at an arbitrary point $\breve{s}$: (a) view showing $\mathcal{H}_{12}$ plane, (b) view orthogonal to $\mathcal{L}\left([\mathcal{H}_{12}, \underline{\mathbf{a}}(\breve{s})]\right)$.

or equivalently, as

$$\sin(\Delta\psi_i) = \frac{\mu_{y_i}}{\hat{\kappa}_1^{-1}} \tag{8.19}$$

$$\Rightarrow \Delta\psi_i = \arcsin(\hat{\kappa}_1 \mu_{y_i}), \quad \text{for } i = 1, 2 \tag{8.20}$$

Hence, substituting $\Delta\psi_i$ given by Eq. (8.19) back into Eq. (8.16) we have

$$\boxed{\begin{aligned}\Delta s_i &= \hat{\kappa}_1^{-1}\Delta\psi_i \\ &= \frac{\arccos(1-\hat{\kappa}_1\mu_{x_i})}{\hat{\kappa}_1} \\ \text{(or, equivalently)} &= \frac{\arcsin(\hat{\kappa}_1\mu_{y_i})}{\hat{\kappa}_1}\end{aligned}} \tag{8.21}$$

It must be noted that if the inclination angle ζ_{inc} (see Eq. (2.55)) of the manifold is 0° then $\zeta = 90°$ and $\hat{\kappa}_1 = \kappa_1$, which is an important feature of symmetric linear arrays.

8.2 Accuracy and the Cramer Rao Lower Bound

The most popular bound in array processing is a well-known statistical result called the Cramer Rao lower bound (CRB) [23]. The CRB sets a lower limit on the error covariance matrix of any unbiased estimate, $\hat{\underline{p}}$, of the true parameter vector $\underline{p} \in \mathcal{R}^M$ in the array signal model of Eq. (1.29). In the case of an array of N sensors receiving M narrowband signals with additive sensor noise of power σ^2, and for a sufficiently large number of snapshots ($L \gg 1$), the expression for the deterministic CRB has been shown to be as follows [24]:

$$\text{CRB}[\underline{p}] = \frac{\sigma^2}{2L}\left(\text{Re}(\mathbb{H}\odot\mathbb{R}_m^T)\right)^{-1} \in \mathcal{R}^{M\times M} \tag{8.22}$$

where

$$\begin{cases}\mathbb{H} = \dot{\mathbb{A}}^H \mathbb{P}_{\mathbb{A}}^{\perp}\dot{\mathbb{A}} \in \mathcal{C}^{M\times M} \\ \mathbb{R}_m^T = \mathcal{E}\{\underline{m}(t)\underline{m}(t)^H\} \in \mathcal{C}^{M\times M} = \text{source covariance matrix}\end{cases} \tag{8.23}$$

with

$$\begin{cases}\mathbb{A} = [\underline{\mathbf{a}}_1, \underline{\mathbf{a}}_2, \ldots, \underline{\mathbf{a}}_M] \\ \dot{\mathbb{A}} = [\dot{\underline{\mathbf{a}}}_1, \dot{\underline{\mathbf{a}}}_2, \ldots, \dot{\underline{\mathbf{a}}}_M] \\ \mathbb{P}_{\mathbb{A}}^{\perp} = \mathbb{I}_N - \mathbb{A}(\mathbb{A}^H\mathbb{A})^{-1}\mathbb{A}^H\end{cases} \tag{8.24}$$

based on the following assumptions:

- $N > M$ and the manifold vectors are independent,
- Noise is a zero-mean, temporally white Gaussian process,

- Noise is spatially white from sensor to sensor; i.e.

$$\mathcal{E}\{\underline{n}(t)\underline{n}^H(t)\} = \sigma^2 \mathbb{I}_N,$$

- Parameters other than $\underline{p}$ are known *a priori.*

In this section the nature of the CRB and its relationship with the differential geometry of the array manifold (and hence the array geometry) is clarified by focusing on the special cases of one emitter as well as two emitters located close together.

8.2.1 *Single Emitter CRB in Terms of Manifold's Differential Geometry*

Assume that the array receives a single signal $m(t)$ of power

$$P = \mathcal{E}\{m(t)m^*(t)\}$$

from bearing p. Then, since $\mathbb{R}_m = P$ and $\mathbb{A} = \underline{\mathbf{a}}(p)$, Eq. (8.22) implies that the CRB may be expressed as

$$\mathrm{CRB}[p] = \frac{\sigma^2}{2LP} \frac{1}{\underline{\dot{\mathbf{a}}}^H(p)\mathbb{P}_{\underline{\mathbf{a}}}^{\perp}\underline{\dot{\mathbf{a}}}(p)} \tag{8.25}$$

Recalling that $\underline{\dot{\mathbf{a}}}(p) = \underline{u}_1(s)\ \dot{s}(p)$ and $\|\underline{u}_1(s)\| = 1$ the above equation becomes

$$\mathrm{CRB}[p] = \frac{\sigma^2}{2LP\dot{s}(p)^2} \frac{1}{\underline{u}_1^H(s)\mathbb{P}_{\underline{\mathbf{a}}}^{\perp}\underline{u}_1(s)} \tag{8.26}$$

However, $\underline{u}_1(s)$ is orthogonal to $\underline{\mathbf{a}}(p)$. Hence $\mathbb{P}_{\underline{\mathbf{a}}}^{\perp}\underline{u}_1(s) = \underline{u}_1(s)$ and Eq. (8.26) may be further simplified as follows:

$$\boxed{\begin{gathered}\mathrm{CRB}[p] = \frac{\sigma^2}{2LP\dot{s}(p)^2} = \frac{1}{2(\mathrm{SNR} \times L)\dot{s}(p)^2} \\ \text{where } \mathrm{SNR} = P/\sigma^2\end{gathered}} \tag{8.27}$$

Note that the above expressions are valid for any manifold curve, not necessarily of hyperhelical shape. For instance Eq. (8.27) can be written so as to provide information about the accuracy of p estimates in terms of the

Table 8.1 Rate-of-change of arc length for linear and planar arrays. The planar arrays are parametrized in terms of (θ, ϕ) and (α, β).

Array	Curve $\mathcal{A}$	Rate of change of arc length s	Parameter p
Linear	$\mathcal{A}$	$\dot{s}(\theta) = \pi \lVert \underline{r}_x \rVert \sin\theta$	$p = \theta$
Planar (θ, ϕ)	$\mathcal{A}_{\phi\vert\theta_o}$	$\dot{s}(\phi) = \pi \lVert \underline{r}(\theta_o) \rVert \sin\phi$	$p = \phi$
	$\mathcal{A}_{\theta\vert\phi_o}$	$\dot{s}(\theta) = \pi \lVert \dot{\underline{r}}(\theta) \rVert \cos\phi_o$	$p = \theta$
Planar (α, β)	$\mathcal{A}_{\alpha\vert\beta_o}$	$\dot{s}(\alpha) = \pi \lVert \underline{r}(\Theta_o) \rVert \sin\alpha$	$p = \alpha$
	$\mathcal{A}_{\beta\vert\alpha_o}$	$\dot{s}(\beta) = \pi \lVert \underline{r}(\Theta_o + 90^\circ) \rVert \sin\beta$	$p = \beta$

sensor locations of a linear array for $p = \theta$, or of a planar array for $p = \theta$, ϕ, α or β by replacing its corresponding $\dot{s}(p)$ as shown in Table 8.1.

For an emitter at a bearing (p, q) the p-estimates correspond to points on the curve $\mathcal{A}_{p|q}$ lying on the manifold surface $\mathcal{M}$ as the parameter q is assumed *constant* and *known*. Thus the notation

$$\mathrm{CRB}[p|\mathcal{A}_{p|q_o}] \tag{8.28}$$

is adopted to represent the CRB for the p-estimates and show its dependence on the properties of the $\mathcal{A}_{p|q_o}$ curve, with the subscript 'o' indicating that q is assumed constant and known. Similarly, the q-estimates are associated with the curve $\mathcal{A}_{q|p}$ on $\mathcal{M}$ as the parameter p is assumed *constant* and *known*, with the CRB of q-estimates represented by

$$\mathrm{CRB}[q|\mathcal{A}_{q|p_o}] \tag{8.29}$$

Using this notation in conjunction with Eq. (8.27) and Table 8.1 for an emitter at a direction (θ, ϕ), the CRB for the ϕ-estimates and θ-estimates are expressed as follows

$$\begin{cases} \mathrm{CRB}[\phi|\mathcal{A}_{\phi|\theta_o}] = \dfrac{1}{2\,(\mathrm{SNR} \times L)\ \left(\pi\, \lVert \underline{r}(\theta_o) \rVert \sin\phi\right)^2} \\ \mathrm{CRB}[\theta|\mathcal{A}_{\theta|\phi_o}] = \dfrac{1}{2\,(\mathrm{SNR} \times L)\ \left(\pi\, \lVert \dot{\underline{r}}(\theta) \rVert \cos\phi_o\right)^2} \end{cases} \tag{8.30}$$

where as usual $\underline{r}(\theta) = \underline{r}_x \cos\theta + \underline{r}_y \sin\theta$ and $\dot{\underline{r}}(\theta) = \frac{d\underline{r}(\theta)}{d\theta}$.

Based on the above discussion a number of observations can be made as follows:

(1) For all planar array geometries:

$$\begin{cases} \mathrm{CRB}[\phi|\mathcal{A}_{\phi|\theta_o}] \propto \dfrac{1}{\sin^2\phi} \\ \mathrm{CRB}[\theta|\mathcal{A}_{\theta|\phi_o}] \propto \dfrac{1}{\cos^2\phi_o} \end{cases} \tag{8.31}$$

The above expressions imply that the θ estimates are more accurate for $\phi_o \to 0°$ while the ϕ estimates are more accurate for $\phi \to 90°$. Furthermore, the variations of $\mathrm{CRB}[\phi|\mathcal{A}_{\phi|\theta_o}]$ and $\mathrm{CRB}[\theta|\mathcal{A}_{\theta|\phi_o}]$ with respect to elevation ϕ are both independent of the array geometry, monotonic and 90° out of phase.

(2) Since $\underline{\dot{r}}(\theta) = \underline{r}(\theta + 90°)$, consequently for all planar array geometries

$$\mathrm{CRB}[\theta|\mathcal{A}_{\theta|\phi_o}] = \mathrm{CRB}[(90° - \phi_o)\,|\mathcal{A}_{\phi|(\theta_o+90°)}] \tag{8.32}$$

In other words, the CRB for the θ-estimates of an emitter at bearings (θ, ϕ) equals the CRB for the ϕ-estimates of a similar emitter at bearing $(\theta + 90°,\ 90° - \phi)$. Furthermore, the variations of $\mathrm{CRB}[\theta|\mathcal{A}_{\theta|\phi_o}]$ and $\mathrm{CRB}[\phi|\mathcal{A}_{\phi|\theta_o}]$ with respect to azimuth θ are both functions of the array geometry, and are 90° out of phase.

(3) For balanced-symmetric arrays

$$\|\underline{\dot{r}}(\theta)\|^2 = \|\underline{r}(\theta)\|^2 = \|\underline{r}_x\|^2 \text{ (independent of } \theta) \tag{8.33}$$

therefore:

$$\begin{cases} CRB[\phi|\mathcal{A}_{\phi|\theta_o}] & = \dfrac{1}{2L \times \mathrm{SNR}\ (\pi\,\|\underline{r}_x\|\sin(\phi))^2} \\ & \Rightarrow \phi \text{ accuracy is independent of } \theta \\ CRB[\theta|\mathcal{A}_{\theta|\phi_o}] & = \dfrac{1}{2L \times \mathrm{SNR}\ (\pi\,\|\underline{r}_x\|\cos(\phi_o))^2} \\ & \Rightarrow \theta \text{ accuracy is independent of } \theta \end{cases} \tag{8.34}$$

Example 8.1 Figure 8.3 shows values of the $\mathrm{CRB}[\theta|\mathcal{A}_{\theta|\phi_o}]$ and $\mathrm{CRB}[\phi|\mathcal{A}_{\phi|\theta_o}]$ for a single emitter at bearing (θ, ϕ), as functions of θ and ϕ. The two equation of Eq. (8.30) are evaluated for the 24-element Y-shaped and circular arrays given in Appendix 8.9.2. The figures are based on the assumption of $L = 100$ snapshots and a Signal-to-Noise ratio of

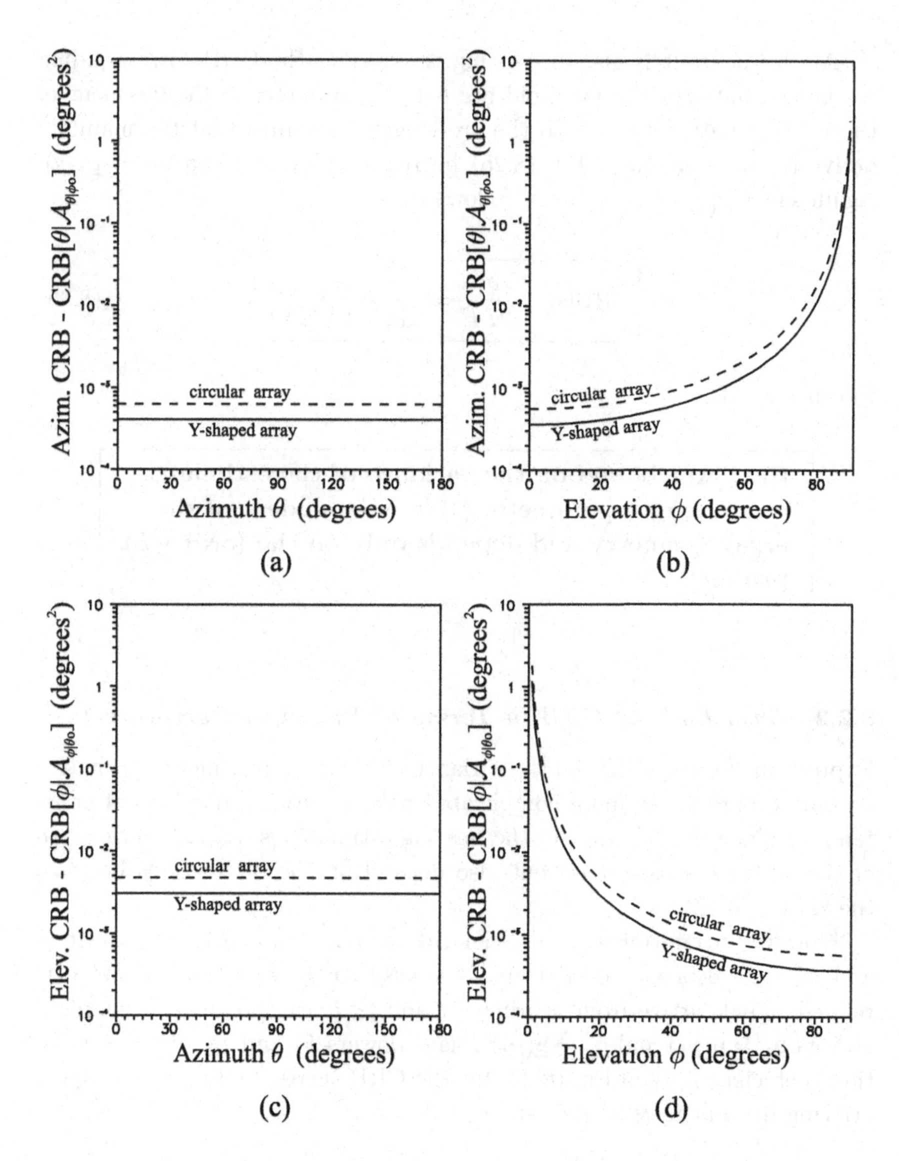

Fig. 8.3 CRB for a single emitter with SNR$\times L = 1000$.

10 dB. Both arrays are balanced-symmetric, and so the accuracy of the parameter estimates will be independent of azimuth. On the other hand, while the accuracy of the parameter estimates varies dramatically with elevation, the performances of the two arrays remain unchanged in relative terms.

At this stage it is also interesting to consider the CRB corresponding to the estimates of the manifold arc length parameter s. Replacement of the bearing parameter p with the arc length s requires that the manifold derivative $\dot{\underline{\mathbf{a}}} = d\underline{\mathbf{a}}/dp$ in Eq. (8.25) be replaced by $\underline{\mathbf{a}}' = d\underline{\mathbf{a}}/ds = \underline{u}_1(s)$. Again since $\mathbb{P}_{\underline{\mathbf{a}}}^{\perp}\underline{u}_1(s) = \underline{u}_1(s)$, it follows that

$$\boxed{\mathrm{CRB}[s] = \frac{\sigma^2}{2LP} = \frac{1}{2(\mathrm{SNR} \times L)}} \tag{8.35}$$

In other words,

> **the lower bound on the variance of the estimates of arc length parameter (s) is independent of the array geometry and depends only on the (SNR $\times$ L) product.**

8.2.2 *Two Emitter CRB in Terms of Principal Curvature*

Expressions for the CRB (on the variance of unbiased parameter estimates) become progressively more complicated with increasing numbers of emitters, M, since the accuracy of the bearing estimates is not only a function of the additive sensor noise but also depend on the interactions between the various emitters.

Consider a multiple-emitter scenario involving two correlated emitters and $M - 2$ uncorrelated emitters. It is easy to show that if the two correlated signals arrive from bearings p_1 and p_2 (corresponding to manifold vectors $\underline{\mathbf{a}}_1 \triangleq \underline{\mathbf{a}}(p_1)$ and $\underline{\mathbf{a}}_2 \triangleq \underline{\mathbf{a}}(p_2)$), have powers P_1 and P_2, and a correlation coefficient ρ, then Eq. (8.22) for the CRB corresponding to the signal arriving from p_1 may be written as

$$\mathrm{CRB}[p_1] = \frac{1}{2(\mathrm{SNR}_1 \times L)} \frac{1}{\dot{\underline{\mathbf{a}}}_1^H \mathbb{P}_{\mathbb{A}}^{\perp} \dot{\underline{\mathbf{a}}}_1} \frac{1}{1 - \frac{\mathrm{Re}^2[\rho \dot{\underline{\mathbf{a}}}_2^H \mathbb{P}_{\mathbb{A}}^{\perp} \dot{\underline{\mathbf{a}}}_1]}{P_1 P_2 (\dot{\underline{\mathbf{a}}}_1^H \mathbb{P}_{\mathbb{A}}^{\perp} \dot{\underline{\mathbf{a}}}_1)(\dot{\underline{\mathbf{a}}}_2^H \mathbb{P}_{\mathbb{A}}^{\perp} \dot{\underline{\mathbf{a}}}_2)}} \tag{8.36}$$

where $\mathrm{SNR}_1 = P_1/\sigma^2$ and $\mathbb{P}_{\mathbb{A}}^{\perp} = \mathbb{I}_N - \mathbb{A}(\mathbb{A}^H\mathbb{A})\mathbb{A}^H$ with $\mathbb{A}$ being an $(N \times M)$ matrix with columns the M manifold vectors. To make Eq. (8.36) more tractable, consider the scenario where all the M emitters are uncorrelated (emissions from independent sources). Setting $\rho = 0$ in Eq. (8.36) we have

$$\text{CRB}[p_1] = \frac{1}{2(\text{SNR}_1 \times L)} \frac{1}{\underline{\dot{\mathbf{a}}}_1^H \mathbb{P}_{\mathbb{A}}^{\perp} \underline{\dot{\mathbf{a}}}_1} \tag{8.37}$$

The dependence of the CRB on the rate of change of manifold arc length, $\dot{s}(p)$, become apparent by considering $\underline{\mathbf{a}}_1$ and $\underline{\mathbf{a}}_2$ as two points on the same manifold curve $\mathcal{A}$. Then using the expression $\underline{\dot{\mathbf{a}}}(p_1) = \underline{u}_1(s_1)\,\dot{s}(p_1)$, Eq. (8.37) can be written as follows:

$$\text{CRB}[p_1|\mathcal{A}] = \frac{1}{2(\text{SNR}_1 \times L)} \frac{1}{\dot{s}(p_1)^2 \underline{u}_1^H(s_1) \mathbb{P}_{\mathbb{A}}^{\perp} \underline{u}_1(s_1)} \tag{8.38}$$

Further interpretation of Eq. (8.38) in terms of the principal curvature of $\mathcal{A}$ is possible when $M = 2$ (i.e. two-emitter scenario) and the two emitters are closely spaced at bearings p_1 and $p_2 = p_1 + \Delta p$, corresponding to manifold arc lengths $s_1 = \breve{s} - \Delta s/2$ and $s_2 = \breve{s} + \Delta s/2$ respectively. Under such circumstances, circular approximation can be applied to a local neighborhood of $\underline{\mathbf{a}}(\breve{s})$ in order to evaluate the term $\underline{u}_1^H(s_1)\ \mathbb{P}_{\mathbb{A}}^{\perp}\ \underline{u}_1(s_1)$. As shown in the Appendix 8.9.3, this hence leads to the following expression:

$$\boxed{\text{CRB}[p_1|\mathcal{A}] = \frac{1}{(\text{SNR}_1 \times L)} \frac{2}{\dot{s}(p_1)^2\ (\Delta s)^2\ \left(\hat{\kappa}_1^2(\breve{p}) - \frac{1}{N}\right)}} \tag{8.39}$$

$$\text{where} \begin{cases} \Delta s = \dot{s}(\breve{p})\ \Delta p \quad \text{and} \\ \hat{\kappa}_1(p) = \kappa_1(p)\ \sin(\zeta(p)) \end{cases} \tag{8.40}$$

Note that the bearing $\breve{p}$ corresponds to the point with arc length $\breve{s}$, and to a first-order approximation, equals $(p_1 + p_2)/2$ or equivalently $\breve{p} = p_1 + \frac{\Delta p}{2}$.

It is important to point out that, since $\Delta s = \dot{s}(\breve{p})\Delta p$, Eq. (8.39) indicates that the CRB is inversely proportional to the square of the emitters' angular separation,

$$\text{CRB}[p_1|\mathcal{A}] \propto \frac{1}{(\Delta p)^2} \tag{8.41}$$

i.e. the accuracy of the system deteriorates rapidly as the emitters approach one another.

The technique deployed for the derivation of Eq. (8.39) can also be directly applied to scenarios involving more than two closely spaced emitters. However, due to the proliferating number of manifold vector inner

Table 8.2 Principal curvature for linear and planar arrays parametrized — planar arrays are in terms of (θ, ϕ) and (α, β).

Array	Curve $\mathcal{A}$	Principal curvature κ_1	Hyperhelix
Linear	A	$\kappa_1 = \|\tilde{\underline{r}}_x^2\|, \forall s$	Yes
Planar (θ, ϕ)	$\mathcal{A}_{\phi\|\theta_o}$	$\kappa_1 = \|\tilde{\underline{r}}^2(\theta_o)\|, \forall s$	Yes
	$\mathcal{A}_{\theta\|\phi_o}$	$\kappa_1(\theta) = \underbrace{\frac{1}{\|\dot{\underline{r}}(\theta)\|^2} \left\| \dot{\underline{r}}^2(\theta) + \frac{j}{\pi \cos\phi_o} P^{\perp}_{\dot{\underline{r}}(\theta)} \underline{r}(\theta) \right\|}_{\simeq \|\widetilde{\dot{\underline{r}}}(\theta)\|^2}$	No
Planar (α, β)	$\mathcal{A}_{\alpha\|\beta_o}$	$\kappa_1 = \|\tilde{\underline{r}}^2(\Theta_o)\|, \forall s$	Yes
	$\mathcal{A}_{\beta\|\alpha_o}$	$\kappa_1 = \|\underline{r}^2(\Theta_o + 90^\circ)\|$	Yes
		where $\widetilde{\underline{R}} = \frac{\underline{R}}{\|\underline{R}\|}$ and $\dot{\underline{R}} = \frac{d\underline{R}}{dp}$	

products, the expression for the CRB can become very cumbersome. Nevertheless, it can be shown that so long as the additional manifold points $s_3, s_4, \ldots, s_M$ are not in the neighborhood of $\breve{s}$ (i.e. in the neighborhood of s_1 and s_2), the value of $\text{CRB}[p_1|\mathcal{A}]$ is primarily dominated by the presence of the source at bearing p_2.

Example 8.2 In Fig. 8.4 the value of $\text{CRB}_\theta[\theta_1]$, as defined in [24], is compared to that given by Eq. (8.39) for two unit-power emitters located at bearings $\theta_1 = 20^\circ$ and $\theta_2 = \theta_1 + \Delta\theta$, and both at a common elevation of $\phi = 20^\circ$. The expressions are evaluated for the 24-element Y-shaped array, assuming the availability of $L = 100$ snapshots and a signal-to-noise ratio of 10 dB (for each emitter).

As can be seen, for small emitter separations, Eq. (8.39) provides an excellent estimate of the CRB. However for increasing separations the circular approximation breaks down and Eq. (8.39) is no longer valid. Note that for large emitter separations, the exact CRB settles down to the value corresponding to a single emitter.

8.2.2.1 *Elevation Dependence of Two Emitters' CRB*

For a planar array with two emitters on the same ϕ-curve $\mathcal{A}_{\phi|\theta_o}$, the dependence of the CRB on elevation ϕ may be deduced by recalling from Table 8.1 (page 181) that

$$\dot{s}_\theta(\theta) \propto \cos\phi \quad \text{and} \quad \dot{s}_\phi(\phi) \propto \sin\phi \tag{8.42}$$

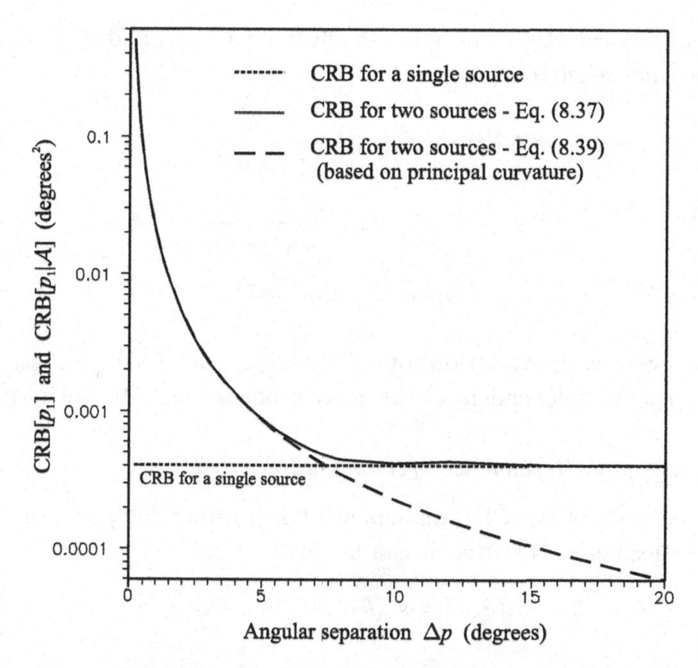

Fig. 8.4 Plots of the CRB Eqs. (8.37) (exact) and (8.39) (appr.) for two emitters with angular separation $\Delta\theta$ (Y-shaped array of 24 sensors). The CRB for a single source is also shown for comparison.

where the subscripts θ and ϕ have been added for reference to θ-curves and ϕ-curves respectively.

Also since the ϕ-curves are hyperhelical, $\hat{\kappa}_{1,\phi}(\phi)$ is independent of ϕ. Furthermore, for large arrays, $\hat{\kappa}_{1,\theta}(\theta)$ is independent of ϕ so long as elevation is not close to 90°. The combination of these results, along with Eq. (8.39) reveal the following dependencies on elevation ϕ:

- For emitters corresponding to two points on the same θ-curve $\mathcal{A}_{\theta|\phi_o}$, that is having the same elevation angle, i.e. (θ_1,ϕ_o) and (θ_2,ϕ_o),

$$\text{CRB}[\theta_1|\mathcal{A}_{\theta|\phi_o}] \propto \frac{1}{\dot{s}_\theta^2(\theta_1)\dot{s}_\theta^2(\breve{\theta})} \tag{8.43}$$

$$\propto \frac{1}{\cos^4\phi_o} \tag{8.44}$$

$$\text{where} \quad \breve{\theta} = \frac{\theta_1+\theta_2}{2} \tag{8.45}$$

- For emitters with the same azimuth angle, i.e. (θ_o,ϕ_1) and (θ_o,ϕ_2), hence on the same ϕ-curve $\mathcal{A}_{\phi|\theta_o}$

$$\mathrm{CRB}[\phi_1|\mathcal{A}_{\phi|\theta_o}] \propto \frac{1}{\dot{s}_\phi^2(\phi_1)\dot{s}_\phi^2(\breve{\phi})} \tag{8.46}$$

$$\propto \frac{1}{\sin^2\phi_1 \sin^2\breve{\phi}} \tag{8.47}$$

$$\text{where} \quad \breve{\phi} = \frac{\phi_1+\phi_2}{2} \tag{8.48}$$

Consequently, the variations of $\mathrm{CRB}[\theta_1|\mathcal{A}_{\theta|\phi_o}]$ and $\mathrm{CRB}[\phi_1|\mathcal{A}_{\phi|\theta_o}]$ with respect to ϕ are independent of the array geometry and 90° out of phase.

8.2.2.2 *Azimuth Dependence of Two Emitters' CRB*

The dependence of the CRB on azimuth θ is a rather complex function of the array geometry. However, it can be proven that

$$\dot{s}_\theta(\theta,\phi) = \dot{s}_\phi(\theta+90°, 90°-\phi) \tag{8.49}$$

and

$$\hat{\kappa}_{1\theta}(\theta,\phi) \simeq \hat{\kappa}_{1\phi}(\theta+90°,\phi) = (\text{independent of } \phi) \tag{8.50}$$

Incorporating these results into Eq. (8.39), and on the grounds that for closely spaced emitters

$$\dot{s}_\theta(\theta_1) \simeq \dot{s}_\theta(\breve{\theta}) \quad \text{and} \quad \dot{s}_\phi(\phi_1) \simeq \dot{s}_\phi(\breve{\phi}), \tag{8.51}$$

one may deduce that:

- if $\Delta\theta = \Delta\phi$, then the CRB on the azimuth estimates of two emitters equally distributed about azimuth $\breve{\theta}$ and with common elevations $\breve{\phi}$, is equal to the CRB on the elevation estimates of two similar emitters equally distributed about elevation $(90° - \breve{\phi})$ and with common azimuth $\breve{\theta} + 90°$. This is somewhat similar to the result derived for the single-emitter case (see Eq. (8.32)) in that the variations of $\mathrm{CRB}[\theta_1|\mathcal{A}_{\theta|\phi_o}]$ and $\mathrm{CRB}[\phi_1|\mathcal{A}_{\phi|\theta_o}]$ with respect to θ are 90° out of phase.

Example 8.3 Figures 8.5(a) and (b) show values of $\mathrm{CRB}[\theta_1|\mathcal{A}_{\theta|\phi}]$ (Eq. (8.39)) as a function of θ and ϕ, in the case of two unit-power emitters of common elevation ϕ but separated by $\Delta\theta = 1°$ about azimuth θ. The bound is evaluated for the 24-element Y-shaped and circular arrays assuming the availability of $L = 100$ snapshots and signal-to-noise ratio

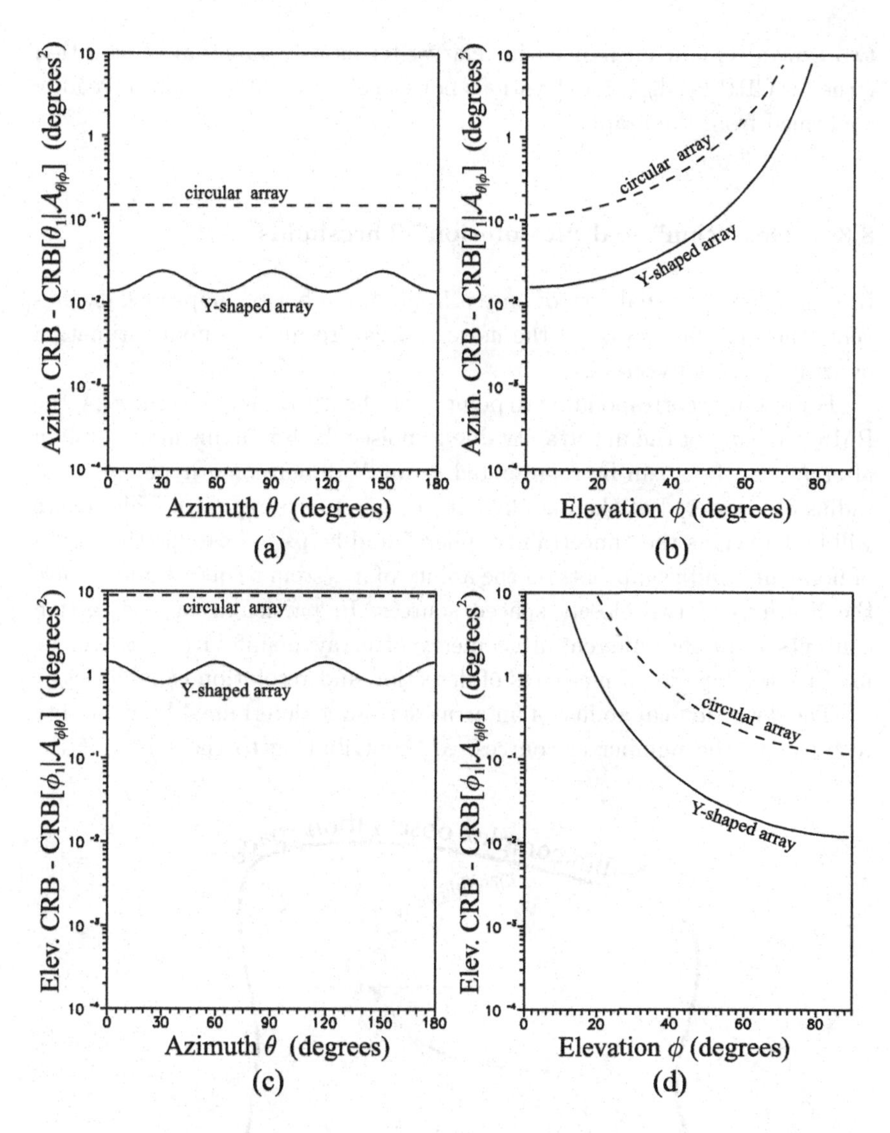

Fig. 8.5 CRL for two emitters with SNR$\times L = 1000$.

of 10 dB (for each emitter). In Fig. 8.5(c) and (d) the exercise is repeated for CRB$[\phi_1|\mathcal{A}_{\phi|\theta}]$ with emitters of common azimuth θ but separated by $\Delta\phi = 1°$ about elevation ϕ. While the circular array again exhibits uniform accuracy for all azimuths, the Y-shaped array shows fluctuations with a period of 60° due to its special shape. Again, the performances of the two arrays remain unchanged in relative terms as a function of elevation (true for $\text{CRB}[\theta_1|\mathcal{A}_{\theta|\phi}]$ for elevations not too close to 90°) as can be readily confirmed from the graphs.

8.3 "Detection" and "Resolution" Thresholds

In practice, the availability of only a limited number of snapshots, L, prevents the full elimination of the noise and can result in a poor estimation of array manifold vectors.

For a source corresponding to point s_i on the array manifold curve $\mathcal{A}$, the RMS value σ_{e_i} of the uncertainty due to noise which remains in the system after L snapshots can be represented as an N-dimensional hypersphere of radius σ_{e_i} centered at the manifold vector $\underline{\mathbf{a}}(s_i)$ (see Fig. 8.6). This sphere will be known as the "uncertainty sphere" and helps to examine the effects of noise and finite snapshots on the ability of a system to detect and resolve the bearings of two closely spaced sources. In particular in this section concepts from the differential geometry of array manifold curves will be used to develop global measures of detection and resolution capabilities.

The detection capability of an array system is determined by its ability to estimate the number of sources, M, contributing to the signal at the

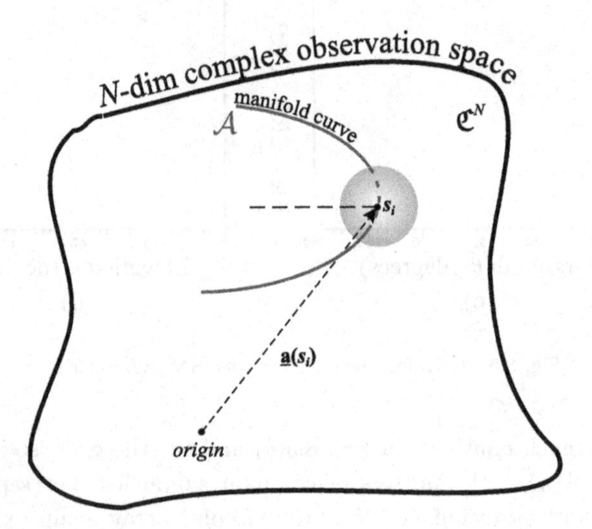

Fig. 8.6 View of the $\mathcal{H}_{12}$ subspace (plane): An illustrative representation of the geometry at '*resolution threshold*' (circular approximation)

array output. Once this information is available, the parameters of interest can be estimated, provided that all the M sources present are resolved. The array resolution is the ability to obtain distinct, albeit inaccurate, parameter estimates for each individual source present. In other words, the sources present are resolved when the corresponding points on the array manifold can be estimated (this, for instance, may correspond to distinct peaks- or nulls-observed in the spectrum of the DF algorithm employed).

However both problems become very difficult if there are sources located "close together" and the determination of "how close" two sources can be, provides the *ultimate* detection and resolution capabilities of the array. In order to analytically determine these ultimate array capabilities it is first necessary to define the detection and resolution "threshold subspaces" (hyperplanes) and then to provide the conditions under which the "thresholds" of *detection* and *resolution* occur. Thus consider two closely located sources of powers P_1 and P_2 arriving from bearings p_1 and $p_2 = p_1 + \Delta p$ corresponding to the manifold points s_1 and $s_2 = s_1 + \Delta s$ respectively. The detection subspace $\mathcal{H}_{\text{det}}$ and resolution subspaces $\mathcal{H}_{\text{res}}$ are defined for a point $\breve{s}$ between s_1 and s_2 as follows:

$$\text{detection:} \quad \mathcal{H}_{\text{det}} \triangleq \mathcal{L}\left([\underline{\mathbf{a}}(\breve{s}),\, \mathbb{P}_{\mathbb{A}}\underline{\mathbf{a}}(\breve{s})]\right) \tag{8.52}$$

$$\text{resolution:} \quad \mathcal{H}_{\text{res}} \triangleq \mathcal{L}\left([\underline{\mathbf{a}}(\breve{s}),\, \underline{u}_1(\breve{s})]\right) \tag{8.53}$$

where

$$\mathbb{P}_{\mathbb{A}} = \mathbb{A}(\mathbb{A}^H\mathbb{A})^{-1}\mathbb{A}^H \quad \text{with} \quad \mathbb{A} = [\underline{\mathbf{a}}(s_1), \underline{\mathbf{a}}(s_2)] \tag{8.54}$$

These two subspace are illustrated in Figs. 8.7 and 8.8 together with the two manifold vectors $\underline{\mathbf{a}}(s_1)$ and $\underline{\mathbf{a}}(s_2)$, and their associated uncertainty spheres. Note that in these figures the manifold curve may be the manifold of a linear array or a p-curve lying on the manifold surface of a planar array.

8.3.1 *Estimating the Detection Threshold*

Two sources corresponding to points s_1 and $s_2 = s_1 + \Delta s$ on the array manifold $\mathcal{A}$ are detected if and only if the uncertainty spheres do not make contact with the detection threshold subspace $\mathcal{H}_{\text{det}}$ defined in Eq. (8.52). When, however, the subspace $\mathcal{H}_{\text{det}}$ becomes the tangent plane to the uncertainties spheres *the threshold of detection occurs.* To a first order approximation, this is equivalent to the following definition.

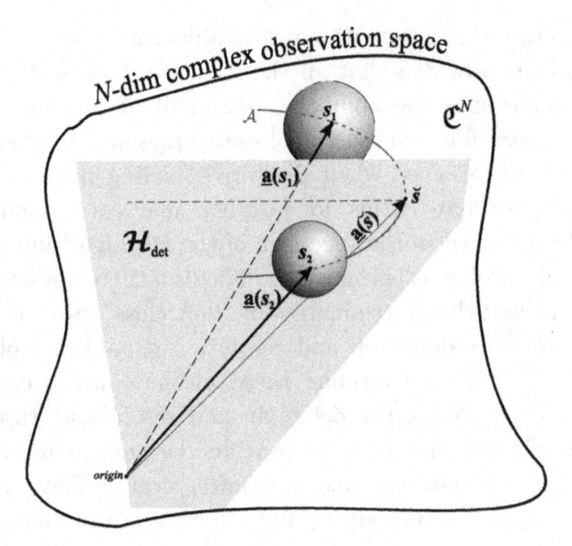

Fig. 8.7 Detection threshold subspace — an illustrative visualization.

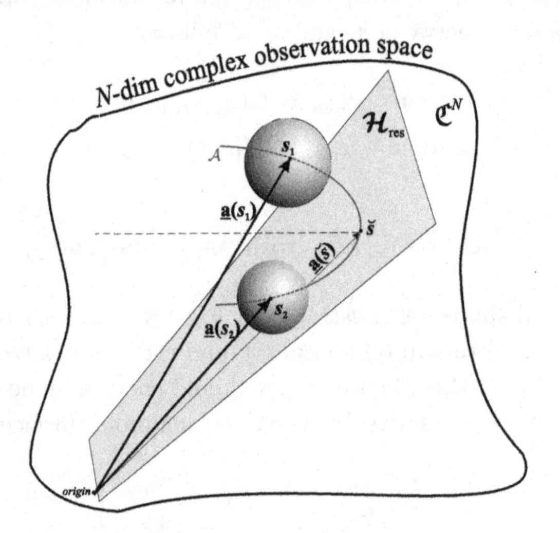

Fig. 8.8 Resolution threshold subspace — an illustrative visualization.

Definition 8.1 Detection Threshold: Two sources are detected if and only if the uncertainty spheres do not make contact. The threshold of detection occurs when the two uncertainty spheres just make contact.

The two sources remain undetectable if their associated uncertainty spheres overlap. This implies that the arc length separation $\Delta s = |s_2 - s_1|$ between two points s_1 and s_2 associated with two sources should be greater

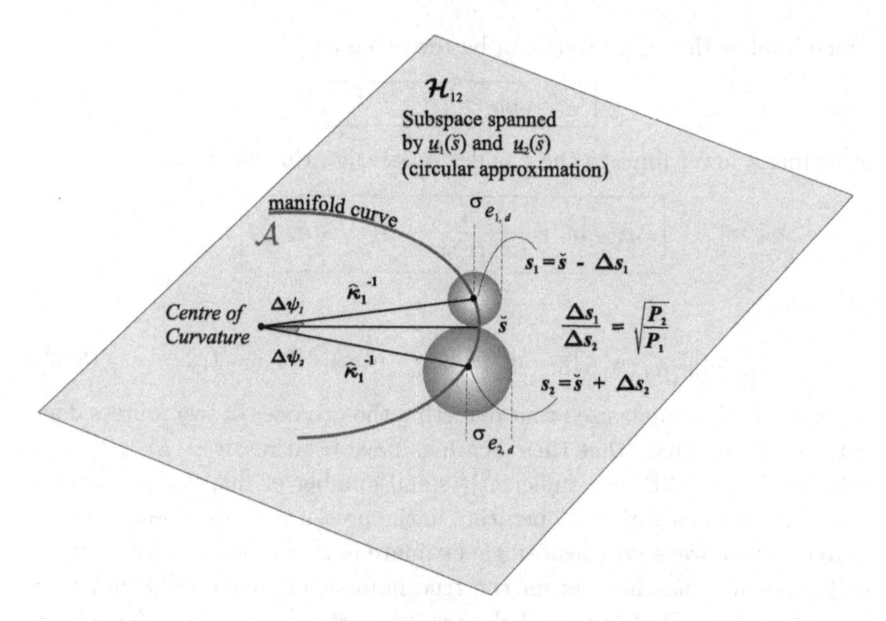

Fig. 8.9 An illustrative representation of the geometry at "*detection threshold*" on the $\mathcal{H}_{12}$ subspace (circular approximation).

than, or equal to, $\Delta s_{\text{det-thr}}$ in order to be detected. That is

$$\Delta s \geqslant \Delta s_{\text{det-thr}} \tag{8.55}$$

Based on this definition and once again using *the circular approximation/representation* of the array manifold, the scenario of Fig. 8.2, at the detection threshold, pertains:

$$\boxed{\mu_{y_1} = \sigma_{e_{1,d}} \quad \text{and} \quad \mu_{y_2} = \sigma_{e_{2,d}}} \tag{8.56}$$

as shown in Fig. 8.9. Hence Eq. (8.21) becomes

$$\begin{aligned} \Delta s_{\text{det-thr}} &= \hat{\kappa}_1^{-1} \Delta\psi \\ &= \frac{\overbrace{\arcsin(\hat{\kappa}_1 \sigma_{e_{1,d}})}^{\triangleq \Delta\psi_1} + \overbrace{\arcsin(\hat{\kappa}_1 \sigma_{e_{2,d}})}^{\triangleq \Delta\psi_2}}{\hat{\kappa}_1} \end{aligned} \tag{8.57}$$

Noting that $\arcsin(x) \simeq x$ for $x \ll 1$, one may write

$$\Delta\psi_1 = \arcsin(\hat{\kappa}_1 \sigma_{e_{1,d}}) \simeq \hat{\kappa}_1 \sigma_{e_{1,d}} \tag{8.58}$$

$$\Delta\psi_2 = \arcsin(\hat{\kappa}_1 \sigma_{e_{2,d}}) \simeq \hat{\kappa}_1 \sigma_{e_{2,d}} \tag{8.59}$$

which implies that Eq. (8.57) can be rewritten as

$$\boxed{\Delta s_{\text{det-thr}} = \sigma_{e_{1,d}} + \sigma_{e_{2,d}}} \tag{8.60}$$

providing a lower limit to the angular separation threshold as

$$\boxed{\Delta p_{\text{det-thr}} = \frac{1}{\pi \left\|\underline{r}\right\| \sin \breve{p}}(\sigma_{e_{1,d}} + \sigma_{e_{2,d}})} \tag{8.61}$$

where

$$\sigma_{e_{i,d}} = \left\|\mathbb{P}^{\perp}_{\mathcal{H}_{\text{det}}}\underline{\mathbf{a}}(s_i)\right\| = \underline{\mathbf{a}}^H(s_i)\mathbb{P}^{\perp}_{\mathcal{H}_{\text{det}}}\underline{\mathbf{a}}(s_i) \quad \text{for } i = 1, 2 \tag{8.62}$$

It should be emphasized that detecting the presence of two sources does not necessarily mean that their bearings have been resolved. In fact for a sufficiently low SNR or a sufficiently small number of snapshots, a typical super-resolution algorithm, operating in the presence of two closely located sources, will always provide a single estimate (e.g. a spectrum with a single null) even if it has been given the true number of sources *a priori*, that is equal to two. That is even if the two sources are assumed to have been detected.

8.3.2 *Estimating the Resolution Threshold*

Definition 8.2 Resolution Threshold: Two emitters corresponding to points s_1 and s_2 on the manifold are resolved if and only if the uncertainty spheres around these two points do not make contact with the resolution threshold subspace, $\mathcal{H}_{\text{res}} \triangleq \mathcal{L}([\underline{\mathbf{a}}(\breve{s}), \underline{u}_1(\breve{s})])$. The threshold of resolution $\Delta s_{\text{res-thr}}$ occurs when $\mathcal{H}_{\text{res}}$ is a tangent plane to the uncertainty spheres (i.e. the uncertainty spheres just make contact with the subspace $\mathcal{H}_{\text{res}}$).

This implies that the arc length separation $\Delta s = |s_2 - s_1|$ between two points s_1 and s_2, corresponding to two already detected sources, should be greater than, or equal to, $\Delta s_{\text{res-thr}}$ in order to be resolved. That is

$$\Delta s \geqslant \Delta s_{\text{det-thr}} \tag{8.63}$$

or remain unresolvable otherwise.

The objective here is to estimate the resolution threshold $\Delta s_{\text{res-thr}}$. That is, to determine the minimum Δs which is just sufficient to allow the uncertainty spheres to make contact with the subspace $\mathcal{H}_{\text{res}} \triangleq \mathcal{L}([\underline{u}_1(\breve{s}), \underline{\mathbf{a}}(\breve{s})])$

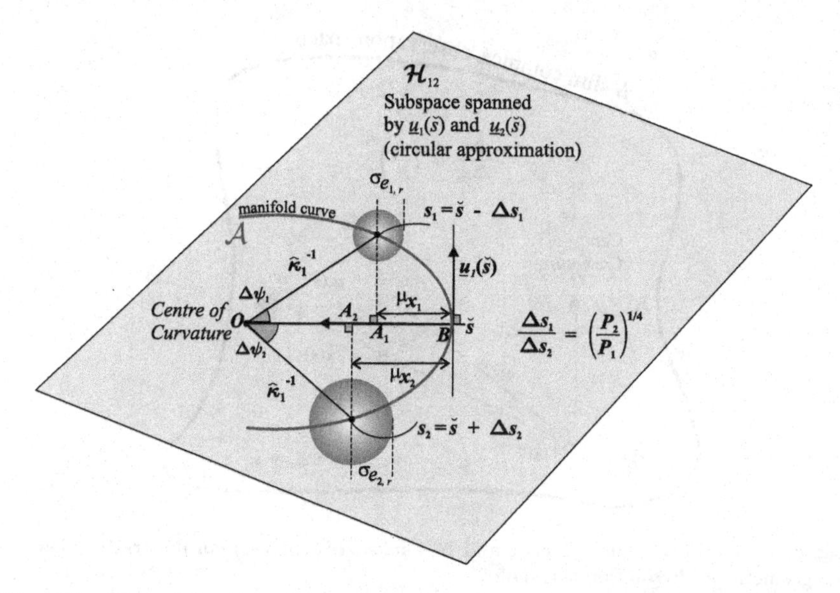

Fig. 8.10 View of the $\mathcal{H}_{12}$ subspace (plane): an illustrative representation of the geometry at "*resolution threshold*" on the $\mathcal{H}_{12}$ subspace (circular approximation).

at the point $\check{s}$. Once again using the circular representation of the array manifold the situation of Definition 8.2 may be illustrated as in Figs. 8.10 and 8.11. With reference to these two figures, it can be seen that $\sigma_{e_{2,r}}$ (see Fig. 8.11) and μ_{x_2} (see Fig. 8.10) as well as $\sigma_{e_{1,r}}$ and μ_{x_1} may be given as follows:

$$\sigma_{e_{1,r}} = \|\mathbb{P}^{\perp}_{\mathcal{H}_{\text{res}}} \underline{\mathbf{a}}(s_1)\| \quad \text{and} \quad \sigma_{e_{2,r}} = \|\mathbb{P}^{\perp}_{\mathcal{H}_{\text{res}}} \underline{\mathbf{a}}(s_2)\| \tag{8.64}$$

$$\mu_{x_1} = \frac{\sigma_{e_{1,r}}}{\sin\gamma} \quad \text{and} \quad \mu_{x_2} = \frac{\sigma_{e_{2,r}}}{\sin\gamma} \tag{8.65}$$

Hence Eq. (8.21) becomes

$$\Delta s_{\text{res-thr}} = \Delta s_1 + \Delta s_2 = \hat{\kappa}_1^{-1} \Delta\psi$$

$$= \frac{\overbrace{\arccos\left(1 - \hat{\kappa}_1 \frac{\sigma_{e_{1,r}}}{\sin\gamma}\right)}^{\triangleq \Delta\psi_1} + \overbrace{\arccos\left(1 - \hat{\kappa}_1 \frac{\sigma_{e_{2,r}}}{\sin\gamma}\right)}^{\triangleq \Delta\psi_2}}{\hat{\kappa}_1} \tag{8.66}$$

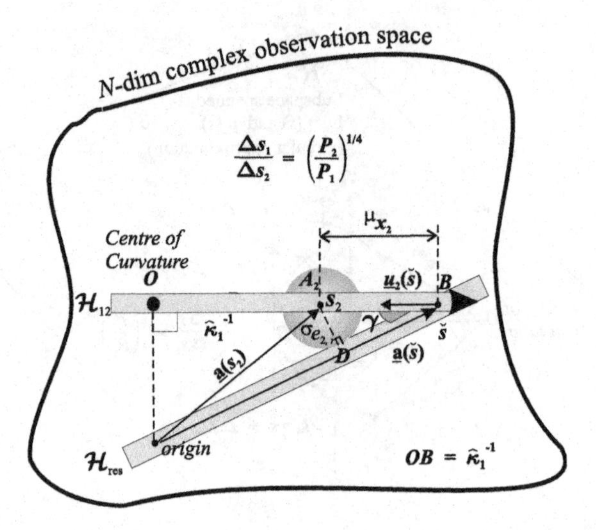

Fig. 8.11 View orthogonal to $\mathcal{H}_{\text{res}}$ and $\mathcal{H}_{12}$ subspaces (planes): an illustrative view of the geometry at "resolution threshold".

Recalling that $\arccos(1 - \frac{1}{2}x^2) \simeq x$ for $x \ll 1$, one may write

$$\begin{aligned}\Delta\psi_1 &= \arccos\left(1 - \hat{\kappa}_1 \frac{\sigma_{e_1}}{\sin\gamma}\right) \\ &\simeq \sqrt{2\hat{\kappa}_1 \frac{\sigma_{e_1}}{\sin\gamma}}\end{aligned} \tag{8.67a}$$

$$\begin{aligned}\Delta\psi_2 &= \arccos\left(1 - \hat{\kappa}_1 \frac{\sigma_{e_2}}{\sin\gamma}\right) \\ &\simeq \sqrt{2\hat{\kappa}_1 \frac{\sigma_{e_2}}{\sin\gamma}}\end{aligned} \tag{8.67b}$$

Upon substitution of Eq. (8.67) into (8.66), it follows that

$$\Delta s_{\text{res-thr}} = \sqrt{\frac{2}{\hat{\kappa}_1 \sin\gamma}} \left(\sqrt{\sigma_{e_{1,r}}} + \sqrt{\sigma_{e_{2,r}}}\right) \tag{8.68}$$

However, (see Fig. 8.11)

$$\cos\gamma = \frac{\|\mathbb{P}^{\perp}_{\mathcal{H}_{1,2}} \underline{\mathbf{a}}(\breve{s})\|}{\|\underline{\mathbf{a}}(\breve{s})\|} = \frac{1/\hat{\kappa}_1}{\sqrt{N}} \Rightarrow \sin^2\gamma = 1 - \cos^2\gamma = 1 - \frac{1}{N\hat{\kappa}_1^2}$$

where

$$\mathcal{H}_{1,2} = \mathcal{L}\{[\underline{u}_1(\check{s}), \underline{u}_2(\check{s})]\}$$

Therefore,

$$\boxed{\Delta s_{\text{res-thr}} = \sqrt[4]{\frac{4}{\hat{\kappa}_1^2 - \frac{1}{N}}} \left(\sqrt{\sigma_{e_{1,r}}} + \sqrt{\sigma_{e_{2,r}}}\right)} \tag{8.69}$$

or equivalently, the lower limit to the angular separation is

$$\boxed{\Delta p_{\text{res-thr}} = \frac{1}{\pi \|\underline{r}\| \sin \check{p}} \underbrace{\sqrt[4]{\frac{4}{\hat{\kappa}_1^2 - \frac{1}{N}}} (\sqrt{\sigma_{e_{1,r}}} + \sqrt{\sigma_{e_{2,r}}})}_{=\Delta s_{\text{res-thr}}}} \tag{8.70}$$

where

$$\sigma_{e_{i,r}} = \left\|\mathbb{P}^{\perp}_{\mathcal{H}_{\text{res}}} \underline{\mathbf{a}}(s_i)\right\| = \underline{\mathbf{a}}^H(s_i) \mathbb{P}^{\perp}_{\mathcal{H}_{\text{res}}} \underline{\mathbf{a}}(s_i), \quad \text{for } i = 1, 2 \tag{8.71}$$

Having defined the required performance thresholds and summarizing the results in Table 8.3, it is apparent that the array detection–resolution capabilities at p are functions of:

- the manifold arc length Δs corresponding to the interval Δp
- the manifold curvatures κ_1 corresponding to bearing p

Table 8.3 Resolution and detection arc length as a function of uncertainty spheres.

Resolution and detection thresholds	
arc length: Δs	directional: $\Delta p \simeq \frac{\Delta s}{\dot{s}(p_o)}$ with $\dot{s}(p_o) = \pi \|\underline{r}\| \sin \check{p}$
$\Delta s_{\text{det-thr}} = \sigma_{e_{1,d}} + \sigma_{e_{2,d}}$	$\Delta p_{\text{det-thr}} \approx \frac{1}{\pi\|\underline{r}\| \sin \check{p}} (\sigma_{e_{1,r}} + \sigma_{e_{2,r}})$
$\Delta s_{\text{res-thr}} = \sqrt[4]{\frac{4}{\hat{\kappa}_1^2 - \frac{1}{N}}} \times \left(\sqrt{\sigma_{e_{1,d}}} + \sqrt{\sigma_{e_{2,d}}}\right)$	$\Delta p_{\text{res-thr}} = \sqrt[4]{\frac{4}{\hat{\kappa}_1^2 - \frac{1}{N}}} \frac{1}{\pi\|\underline{r}\| \sin \check{p}} \times \left(\sqrt{\sigma_{e_{1,r}}} + \sqrt{\sigma_{e_{2,r}}}\right)$
with $\begin{cases} \sigma_{e_{i,d}} = \|P_{\mathcal{H}_{\text{det}}} \underline{\mathbf{a}}(s_i)\|, & i = 1, 2 \\ \sigma_{e_{i,r}} = \|P^{\perp}_{\mathcal{H}_{\text{res}}} \underline{\mathbf{a}}(s_i)\|, & i = 1, 2 \end{cases}$	

- the number of sensors N
- the radius of the uncertainty spheres σ_e

where the fourth factor is dependent on the available SNR and the number of array output snapshots L, and will be modelled next.

8.4 Modelling of the Uncertainty Sphere

As described above, the uncertainty spheres (in arc lengths) represent the effects of additive sensor noise of power σ^2 on the system performance. Furthermore, as was noted earlier, it is possible to asymptotically eliminate the effects of noise by increasing the number of snapshots L. That is

$$\sigma_e^2 \propto \frac{1}{L} \tag{8.72}$$

and as $L \rightarrow \infty \Rightarrow \sigma_e \rightarrow 0$, i.e. zero estimation error. In addition, as the power of the noise in the system tends to zero ($\sigma \rightarrow 0$), i.e. the SNR $= P/\sigma^2$ tends to infinity (SNR $\rightarrow \infty$), the uncertainty sphere shrinks with an uncertainty radius $\sigma_e \rightarrow 0$. That is

$$\sigma_e^2 \propto \frac{1}{\text{SNR}} \tag{8.73}$$

This implies that the uncertainty sphere shrinks, or equivalently, that there would be a gradual decrease in its radius σ_e, thus providing an effective radius σ_e according to the following model:

$$\sigma_e = \sqrt{\frac{\sigma^2}{2LP}} = \sqrt{\frac{1}{2(\text{SNR} \times L)}}; \quad \text{(in arc lengths)} \tag{8.74}$$

It should be pointed out that the factor "2" appearing in the denominator is a direct result of using complex numbers to represent signal envelopes in the array model.

The shrinking rate of the uncertainty sphere depends also on the estimation algorithm deployed. Some algorithms exhibit better properties than others thereby "masking" the overall array performance and this makes it difficult to identify the full detection-resolution capabilities inherent in the array structure. Thus we may introduce the factor C (with $0 < C \leq 1$) to represent the reduction in performance due to the employment of a specific practical estimation algorithm. That is

$$\sigma_e = \sqrt{\frac{1}{2(\mathrm{SNR} \times L)C}}$$

For $C = 1$ the above expression becomes a "lower bound" and forms a benchmark against which any practical algorithm can be compared. One interpretation of the value $C = 1$ may be that it is a theoretical limit achieved by an "ideal" algorithm which does not introduce extra "uncertainties" and eliminates any dependency which may exist between the various parameters of the received signals (for instance it decorrelates any correlated signals, etc).

It is interesting to note at this point the relationship between the uncertainty sphere model of Eq. (8.74) and the variance of the estimate of the manifold's arc length as predicted by the Cramer Rao lower bound of Eq. (8.25). That is,

$$\sigma_e^2 = \mathrm{CRB}[s] \tag{8.75}$$

The equality of the single source Cramer Rao lower bound with the square of the uncertainty sphere radius can be interpreted as follows:

The uncertainty sphere represents the smallest achievable uncertainty (optimal accuracy) due to the presence of noise after L snapshots, when all the effects of the presence of other sources have been eliminated by an "ideal" parameter estimation algorithm.

8.5 Thresholds in Terms of (SNR × L)

Based on the previous discussion and the model of Eq. (8.74), at the detection and resolution "thresholds" the uncertainty spheres can be expressed as

$$\sigma_{e_{i,d}} = \underline{\mathbf{a}}^H(s_i)\mathbb{P}^{\perp}_{\mathcal{H}_{\mathrm{det}}}\underline{\mathbf{a}}(s_i) = \left.\sqrt{\frac{1}{2(\mathrm{SNR} \times L)C}}\right|_{\mathrm{det}} \tag{8.76a}$$

$$\sigma_{e_{i,r}} = \underline{\mathbf{a}}^H(s_i)\mathbb{P}^{\perp}_{\mathcal{H}_{\mathrm{res}}}\underline{\mathbf{a}}(s_i) = \left.\sqrt{\frac{1}{2(\mathrm{SNR} \times L)C}}\right|_{\mathrm{res}} \tag{8.76b}$$

Substituting Eq. (8.76a) back into Eqs. (8.60) and (8.69), with $C = 1$, it follows that:

$$
\begin{aligned}
&\text{(a) Detection threshold:}\\
&\Delta s_{\text{det-thr}} = \sqrt{\text{CRB}[s_1|\mathcal{A}]} + \sqrt{\text{CRB}[s_2|\mathcal{A}]}\\
&\qquad = \frac{1}{\sqrt{2}}\left(\frac{1}{\sqrt{\text{SNR}_1 \times L}} + \frac{1}{\sqrt{\text{SNR}_2 \times L}}\right)\\
&\qquad = \frac{1}{\sqrt{2(\text{SNR}_1 \times L)}}\left(1 + \sqrt{\frac{P_1}{P_2}}\right)\\
&\text{(b) Resolution threshold:}\\
&\Delta s_{\text{res-thr}} = \sqrt[4]{\frac{4}{\left(\hat{\kappa}_1^2 - \frac{1}{N}\right)}}\\
&\qquad \times \left(\sqrt[4]{\text{CRB}[s_1|\mathcal{A}]} + \sqrt[4]{\text{CRB}[s_2|\mathcal{A}]}\right)\\
&\qquad = \sqrt[4]{\frac{4}{\left(\hat{\kappa}_1^2 - \frac{1}{N}\right)}}\frac{1}{\sqrt[4]{2}}\\
&\qquad \times \left(\frac{1}{\sqrt[4]{\text{SNR}_1 \times L}} + \frac{1}{\sqrt[4]{\text{SNR}_2 \times L}}\right)\\
&\qquad = \sqrt[4]{\frac{2}{(\text{SNR}_1 \times L)\left(\hat{\kappa}_1^2 - \frac{1}{N}\right)}}\left(1 + \sqrt[4]{\frac{P_1}{P_2}}\right)
\end{aligned}
\tag{8.77}
$$

The incremental manifold length Δs may also be written in terms of bearing separation Δp as $\Delta s \simeq \Delta p\, \dot{s}(\breve{p})$ where $\dot{s}(\breve{p}) = \pi\|\underline{r}\| \sin \breve{p}$ is the rate of change of manifold arc length and bearing $\breve{p}$ corresponds to the point $\breve{s}$ on the manifold curve (to a first order approximation — $\breve{p} = (p_1 + p_2)/2$). This implies that the minimum angular separation $\Delta p_{\text{det-thr}}$ required for the detection of two sources of powers P_1 and P_2 may be written as:

$$
\begin{aligned}
&\text{(a) Detection threshold:}\\
&\Delta p_{\text{det-thr}} = \frac{1}{\sqrt{2}\dot{s}(\breve{p})}\left(\frac{1}{\sqrt{\text{SNR}_1 \times L}} + \frac{1}{\sqrt{\text{SNR}_2 \times L}}\right)\\
&\text{(b) Resolution threshold:}\\
&\Delta p_{\text{res-thr}} = \frac{1}{\dot{s}(\breve{p})}\sqrt[4]{\frac{2}{\left(\hat{\kappa}_1^2 - \frac{1}{N}\right)}}\\
&\qquad \times \left(\frac{1}{\sqrt[4]{\text{SNR}_1 \times L}} + \frac{1}{\sqrt[4]{\text{SNR}_2 \times L}}\right)
\end{aligned}
\tag{8.78}
$$

Since $\dot{s}(\breve{p}) = \pi \|\underline{r}\| \sin \breve{p}$, Eq. (8.78) indicates that the resolution capabilities of a linear array (or ELA) are maximum at broadside ($p = 90°$) and zero along the endfire directions ($p = 0°$ or $180°$). Zero resolution along end-fire directions is expected since, due to cylindrical symmetry, two signals at directions which are symmetrical to the array axis cannot be distinguished, irrespective of their angular separation.

For the special case of equi-powered sources ($P_1 = P_2 = P$) the ultimate product (signal-to-noise ratio times L) required to detect and resolve two sources at bearings p_1 and $p_2 = p_1 + \Delta p$ may be written as

$$\begin{aligned} \Delta p_{\text{det-thr}} &\approx \frac{\sqrt{2}}{\dot{s}(\breve{p})} \frac{1}{\sqrt{\text{SNR} \times L}} \\ \Delta p_{\text{res-thr}} &\approx \frac{1}{\dot{s}(\breve{p})} \sqrt[4]{\frac{32}{\left(\hat{\kappa}_1^2 - \frac{1}{N}\right)}} \frac{1}{\sqrt[4]{\text{SNR} \times L}} \end{aligned} \tag{8.79}$$

revealing the following *square-root law for detection threshold*:

$$\boxed{\Delta p_{\text{det-thr}} \propto (\text{SNR} \times L)^{-1/2}} \tag{8.80}$$

and *fourth-root law for resolution*:

$$\boxed{\Delta p_{\text{res-thr}} \propto (\text{SNR} \times L)^{-1/4}} \tag{8.81}$$

In other words the detection and resolution ultimately achievable by a linear array is inversely proportional to the square-root and fourth-root law respectively, of

- the signal-to-noise ratio
- the number of snapshots

Furthermore, in the case of equi-powered sources ($P_1 = P_2$) a relationship between the ultimate resolving and detection capabilities of an array can be determined by dividing $\Delta p_{\text{res-thr}}$ by $\Delta p_{\text{det-thr}}$. That is

$$\frac{\Delta p_{\text{res-thr}}}{\Delta p_{\text{det-thr}}} = \sqrt{\frac{8}{\hat{\kappa}_1^2 - \frac{1}{N}}} \sqrt[4]{\text{SNR} \times L} \tag{8.82}$$

which indicates that resolution is a more demanding operation compared to detection, since the right hand side of the above equation is always greater than unity ($\gg 1$). In addition, Eq. (8.79) reveals that the detection threshold falls more rapidly than the resolution threshold as the number of sensors N increases for a constant $\text{SNR} \times L$.

It is interesting to note that the resolution threshold $\Delta p_{\text{res-thr}}$ of a linear array is a function of $\sin(p)$. This indicates that an algorithm which performs an exhaustive search of the manifold (e.g. MUSIC) should evaluate its cost function along a non-uniform grid of azimuths whose values follow a sinusoidal distribution. Such a non-uniform angular grid corresponds to a uniform arc length grid along the manifold and would allow full use of the array's resolution capabilities.

Finally Eq. (8.77) can be used to derive other useful system requirements. For example, the number of snapshots, $L_{\text{det-thr}}$, required to detect two sources at bearings p_1 and $p_2 = p_1 + \Delta p$ or $L_{\text{res-thr}}$ to resolve them, may be written as shown below.

(a) Detection threshold:

$$L_{\text{det-thr}} = \frac{1}{2\Delta s^2}\left(\frac{1}{\sqrt{\text{SNR}_1}} + \frac{1}{\sqrt{\text{SNR}_2}}\right)^2$$

$$(\text{SNR}_1 \times L)_{\text{det-thr}} = \frac{1}{2\Delta s^2}\left(1 + \sqrt{\frac{P_1}{P_2}}\right)^2$$

(b) Resolution threshold:

$$L_{\text{res-thr}} = \frac{2}{\Delta s^4\left(\hat{\kappa}_1^2 - \frac{1}{N}\right)}\left(\frac{1}{\sqrt[4]{\text{SNR}_1}} + \frac{1}{\sqrt[4]{\text{SNR}_2}}\right)^4$$

$$(\text{SNR}_1 \times L)_{\text{res-thr}} = \frac{2}{\Delta s^4\left(\hat{\kappa}_1^2 - \frac{1}{N}\right)}\left(1 + \sqrt[4]{\frac{P_1}{P_2}}\right)^4$$

$$\text{where}\begin{cases}\Delta s = \pi\|\underline{r}\||\cos p_2 - \cos p_1| \\ \Delta s_1\hat{\kappa}_1 \ll 1, \quad \Delta s_2\hat{\kappa}_1 \ll 1\end{cases} \tag{8.83}$$

Above, the expressions are also provided for the minimum product $(\text{SNR}_1 \times L)$ required for a source of power P_1 to be detected and then resolved in the presence of a source of power P_2 after L snapshots when the two sources are separated by Δs.

Finally Table 8.4 summarizes the main expressions for the special case of two equi-power sources.

Table 8.4 Detection and resolution expressions for equipowered sources ($P_1 = P_2 = P$).

Detection	Resolution
$\Delta p_{\text{det-thr}} \approx \frac{\sqrt{2}}{\dot{s}(\check{p})} \frac{1}{\sqrt{\text{SNR} \times L}}$	$\Delta p_{\text{res-thr}} \approx \frac{1}{\dot{s}(\check{p})} \sqrt[4]{\frac{32}{\left(\kappa_1^2 - \frac{1}{N}\right)}} \frac{1}{\sqrt[4]{\text{SNR} \times L}}$
$(\text{SNR} \times L)_{\text{det-thr}} = \frac{2}{\Delta s^2}$	$(\text{SNR} \times L)_{\text{res,thr}} = \frac{32}{\Delta s^4 \left(\kappa_1^2 - \frac{1}{N}\right)}$
$L_{\text{det-thr}} = \frac{2}{\Delta s^2} \frac{1}{\text{SNR}}$	$L_{\text{res-thr}} = \frac{32}{\Delta s^4 \left(\kappa_1^2 - \frac{1}{N}\right)} \frac{1}{\text{SNR}}$
where $\dot{s}(\check{p}) = \pi \, \|\underline{r}\| \sin \check{p}$	

8.6 Comments

8.6.1 *Schmidt's Definition of Resolution*

An alternative definition to the one presented in the previous section (see Definition 8.2 — page 194) for the resolution threshold was proposed by Schmidt [1] stating that two sources corresponding to points s_1 and s_2 on the manifold are resolved if, and only if, the uncertainty spheres do not make contact with the tangent $\underline{u}_1(\check{s})$ to the manifold at $\check{s}$. Using this definition, which was given without any analysis, the resolution threshold subspace $\mathcal{H}_{\text{res}}$ can be defined as

$$\mathcal{H}_{\text{res}} \triangleq \mathcal{L}\left([\underline{u}_1(\check{s})]\right) \tag{8.84}$$

having dimensionality one. Schmidt's definition, in conjunction with the *circular approximation* of the array manifold is illustrated in Fig. 8.12. Based on this figure, it can be proven using the approach presented in the previous sections that the resolution Δs is given by

$$\Delta s_{\text{res-thr}} = \sqrt[4]{\frac{4}{\kappa_1^2}} \left(\sqrt[4]{\text{CRB}[s_1|\mathcal{A}]} + \sqrt[4]{\text{CRB}[s_2|\mathcal{A}]} \right) \tag{8.85a}$$

$$= \sqrt[4]{\frac{4}{\kappa_1^2} \frac{1}{\sqrt[4]{2}}} \left(\frac{1}{\sqrt[4]{\text{SNR}_1 \times L}} + \frac{1}{\sqrt[4]{\text{SNR}_2 \times L}} \right) \tag{8.85b}$$

$$= \sqrt[4]{\frac{2}{(\text{SNR}_1 \times L)\, \kappa_1^2}} \left(1 + \sqrt[4]{\frac{P_1}{P_2}} \right) \tag{8.85c}$$

The above expression is an approximation to Eq. (8.77) where the term $\hat{\kappa}_1^2 - \frac{1}{N}$ has been replaced by κ_1^2. These two expressions become identical

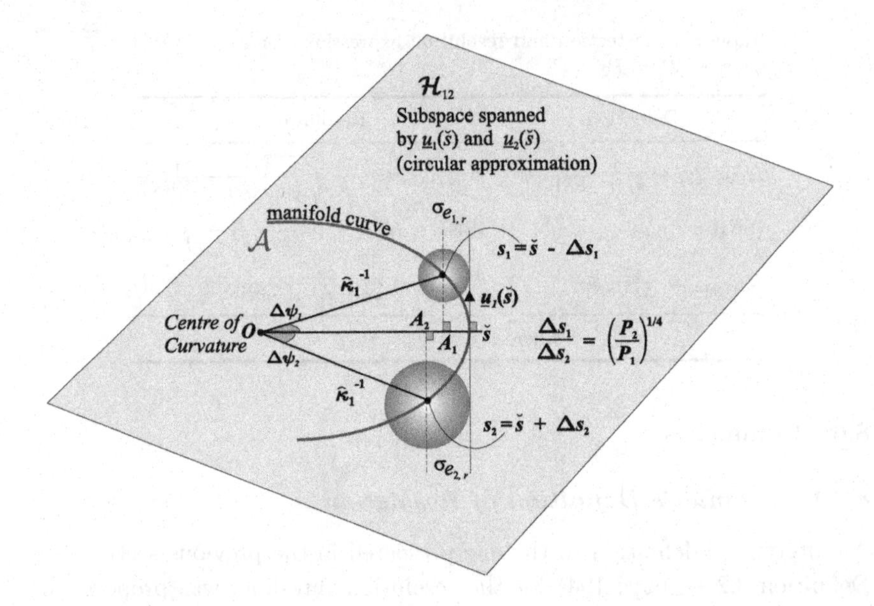

Fig. 8.12 Illustrative representation of Schmidt's definition.

for large N and inclination angle $\zeta = 90°$. That is for symmetric arrays with a large number of sensors.

8.6.2 *CRB at the Resolution Threshold*

If the CRB is evaluated at the threshold level $(\text{SNR}_1 \times L)_{\text{res}}$, then the estimation error (accuracy of the estimate) at the resolution threshold is given as follows:

$$\text{CRB}[p_1|\mathcal{A}]_{\text{res}} = \frac{1}{2(\text{SNR}_1 \times L)_{\text{res}}}(\dot{\underline{\mathbf{a}}}_1^H \mathbb{P} \dot{\underline{\mathbf{a}}}_1)^{-1} \tag{8.86}$$

$$\simeq \frac{1}{\left(1+\sqrt[4]{\frac{P_1}{P_2}}\right)^4} \frac{(\Delta p \times \dot{s}(p))^2}{\dot{s}^2(p_1)} \tag{8.87}$$

It is important to note that for two equi-powered sources the above expression is simplified to

$$\sqrt{\text{CRB}[p_1]_{\text{res}}} \simeq \frac{\Delta p}{4} \cdot \frac{\dot{s}(\breve{p})}{\dot{s}(p_1)}$$

$$\underset{[\text{as } (\breve{p}-p_1)\to 0]}{\longrightarrow} \frac{\Delta p}{4} \tag{8.88}$$

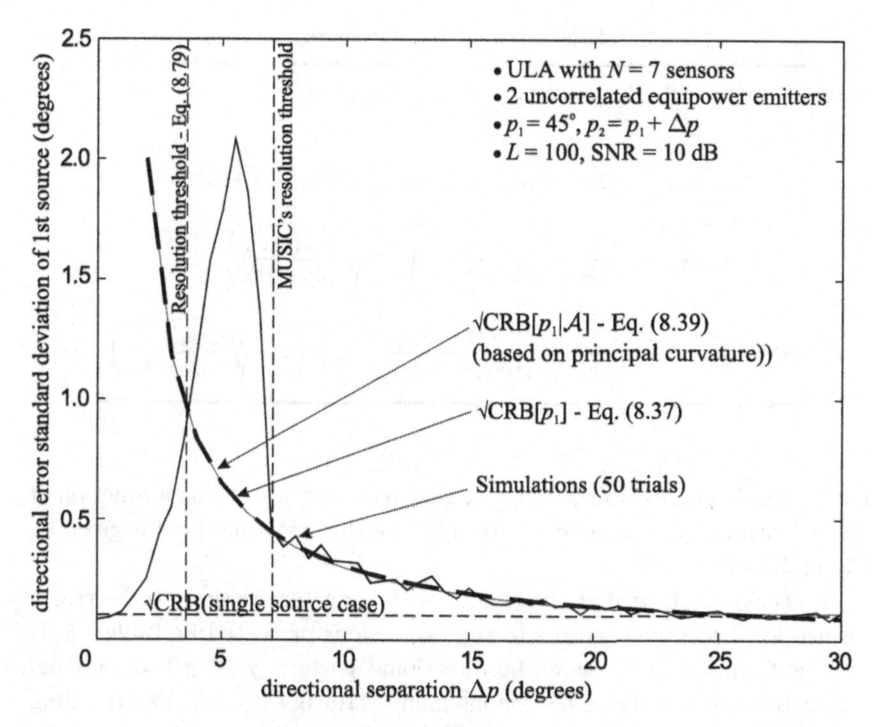

Fig. 8.13 Simulation results of the CRB (exact and approximate expressions) as a function of azimuth separation, averaged over 50 trials. The resolution threshold and the single-source CRB are also shown for comparison.

which is an expected result indicating the generality and significance of the resolution threshold. Figure 8.13 illustrates the variation, with azimuth separation, for the CRB given by Eqs. (8.37) (exact) and (8.39) (approx.) together with the results of simulation studies and the resolution threshold. Remember that for large separations the exact CRB converges to the single-source CRB while the differential geometry version wrongly tends to zero — although this is not obvious in Fig. 8.13.

8.6.3 *Directional Arrays*

The above discussion was carried out under the assumption that the planar array consists of antennas that are isotropic (with gain of unity) in both azimuth θ and elevation ϕ. This assumption might seem unrealistic since many practical antennas (like the elevated-feed monopoles) are

Table 8.5 Expressions for directional array of sensors.

$$\mathrm{CRB}[p_1|\mathcal{A}] = \frac{1}{2\,\mathrm{SNR}_1 \times L|g(p_1)|^2}\,(\dot{\underline{\mathbf{a}}}_1^H\,\mathbb{P}\,\dot{\underline{\mathbf{a}}}_1)^{-1}$$

$$\simeq \frac{1}{(\mathrm{SNR}_1 \times L)|g(p_1)|^2}\,\frac{2}{(\Delta s)_2^2(p_1)(\widehat{\kappa}_1^2(p) - 1/N)}$$

$$(\mathrm{SNR}_1 \times L)_{\mathrm{det}} = \frac{1}{2(\Delta p \times (p))^2|g(p_1)|^2}\left(1 + \sqrt{\frac{|g(p_1)|^2 P_1}{|g(p_2)|^2 P_2}}\right)^2$$

$$(\mathrm{SNR}_1 \times L)_{\mathrm{res}} = \frac{2}{(\Delta p \times (p))^4(\widehat{\kappa}_1^2(p) - \frac{1}{N})|g(p_1)|^2}\left(1 + \sqrt[4]{\frac{|g(p_1)|^2 P_1}{|g(p_2)|^2 P_2}}\right)^4$$

non-isotropic and exhibit a complex gain response $\gamma_k \in \mathfrak{C}^1$ as a function of one or both bearing parameters. In this case the array manifold is given by Eq. (1.22) of Chapter 1.

It is easy to show that the results of the previous section can be transformed for directional sensors to the expressions presented in Table 8.5. It is clear from Table 8.5 that the directional pattern $g(p)$ of a directional sensor behaves simply as a "voltage gain" term boosting or deteriorating the effective Signal-to-Noise Ratio at the output of the array. Consequently, it can be stated that the presence of directional sensors does not affect the relative merits of one array geometry over the other, although of course, performance is affected in absolute terms.

8.7 Array Capabilities Based on α- and β-curves

In the previous sections the estimation accuracy, detection and resolution capabilities of an array of sensors and their dependence on the differential geometry of manifold parameter curves were investigated. It was seen that for sufficiently close emitters, the array performance is a function of the local shape of a manifold curve as specified by its length and first curvature. It was consequently deduced that while performance varies with azimuth θ in accordance with the array configuration, its variation with elevation obeys simple sinusoidal laws irrespective of the array. While these deductions are completely valid, examination of array performance in terms of cone-angles can provide an alternative, and perhaps a clearer picture of the operation of a planar array.

Theorem 8.1 *The capability of a planar array to estimate, detect and resolve cone-angles α_1 and α_2 of rotation angle Θ_o for two emitters at bearings (α_1, β_o) and (α_2, β_o) is*

(i) independent of the value of β_o,
(ii) identical to its capability to estimate, detect and resolve the elevations of two emitters at (azimuth, elevation) bearings:

$$(\theta = \Theta_o, \phi = \alpha_1) \text{ and } (\theta = \Theta_o, \phi = \alpha_2) \quad \text{if } \alpha_1, \alpha_2 < 90^\circ \tag{8.89}$$

or

$$(\theta = \Theta_o, \phi = 180^\circ - \alpha_1) \text{ and } (\theta = \Theta_o, \phi = 180^\circ - \alpha_2) \quad \text{if } \alpha_1, \alpha_2 > 90^\circ \tag{8.90}$$

Proof.

(i) According to Corollary 5.2 and Theorem 5.1, all members of the family of α-curves are identical, i.e. their differential geometry is independent of β.
(ii) From Corollary 5.5 and Theorem 5.1, each member of the α-curve family is identical to the combination of the ϕ-curves corresponding to $\theta_o = \Theta_o$ and $\theta_o = \Theta_o + 180^\circ$. This implies that α-estimation and ϕ-estimation performance must by necessity be equivalent.

□

Naturally, the roles of α and β may be interchanged in Theorem 8.1 by simply replacing Θ_o with $\Theta_o + 90^\circ$.

Part (i) of the above theorem suggests an underlying "conic" behavior in the performance of a planar array which can not be observed by considering azimuth and elevation angles alone. Figure 8.14 illustrates this behavior by showing the loci of two directions of arrival separated by $\Delta\alpha = 2^\circ$ ($\alpha_1 = 68^\circ$, $\alpha_1 = 70^\circ$) for all possible values of β and the corresponding θ-ϕ values.

Part (ii) of Theorem 8.1 indicates that the properties of ϕ-estimation array performance conducted in Section 8.2 is directly applicable to all corresponding cone-angles. In particular, one might immediately deduce that the α- and β-estimation performances of a planar array of omnidirectional sensors are at their peaks when $\alpha = 90^\circ$ and $\beta = 90^\circ$ respectively, and degrade sinusoidally as α and β approach either 0° or 180°.

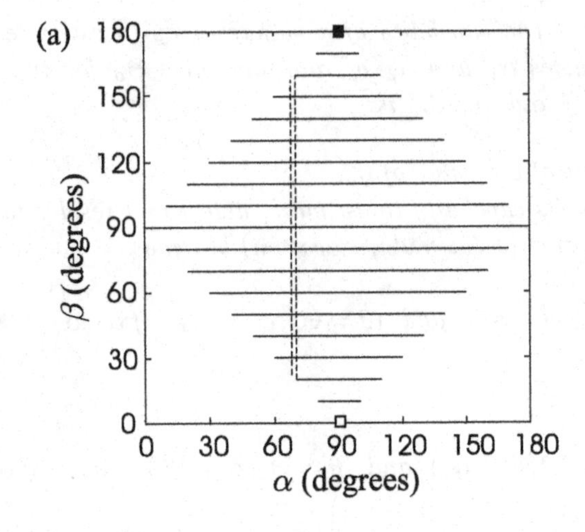

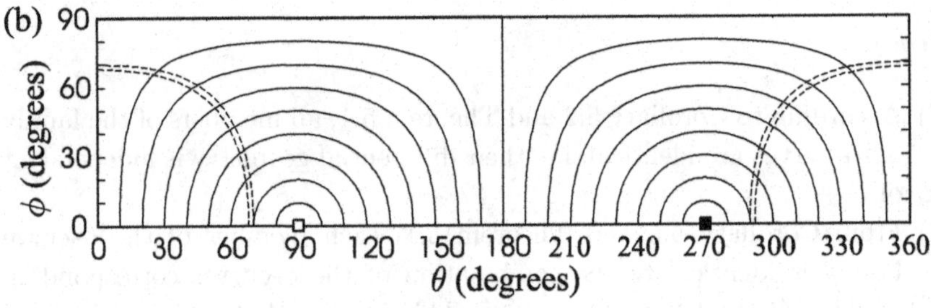

Fig. 8.14 The loci of two DOAs separated by $\Delta\alpha = 2^\circ$ for all values of β and the corresponding DOA loci on the (θ, ϕ) parameter plane revealing the "conic" nature of the array performance.

8.8 Summary

In this chapter resolution and detection thresholds/bounds and estimation accuracy have been studied by approximating, locally, a manifold curve with a circular arc. In particular, by defining the detection and resolution subspaces, in conjunction with the circular approximation of the array manifold, the minimum arc length separations for detecting and resolving two sources located close together have been estimated. This is done in terms of the curve's principal curvature, thereby simplifying the analysis considerably. Furthermore, the relationship between Cramer Rao Lower Bound (CRB) and resolution and detection thresholds has been established. These thresholds were based on differential geometry properties of the manifolds and can be used as a "figure of merit" for comparing array geometries.

8.9 Appendices

8.9.1 *Radius of Circular Approximation*

Based on Fig. 8.2(b) the radius of the corresponding circular arc is

$$\hat{\kappa}_1^{-1} = \|\mathbb{P}_{\mathcal{H}_{12}} \underline{\mathbf{a}}(\breve{s})\| = \sqrt{\underline{\mathbf{a}}^H(\breve{s}) \mathbb{P}_{\mathcal{H}_{12}} \underline{\mathbf{a}}(\breve{s})} \tag{8.91}$$

where $\mathbb{P}_{\mathcal{H}_{12}}$ is the projection operator on to the subspace

$$\mathcal{H}_{12} = \mathcal{L}\left(\left[\underline{u}_1(\breve{s}), \underline{u}_2(\breve{s})\right]\right) \tag{8.92}$$

However,

$$\begin{aligned}
&\underline{\mathbf{a}}^H(\breve{s}) \mathbb{P}_{\mathcal{H}_{12}} \underline{\mathbf{a}}(\breve{s}) \quad \text{(below the variance } \breve{s} \text{ is suppressed for convenience)} \\
&= \underline{\mathbf{a}}^H \left[\underline{u}_1, \ \underline{u}_2\right] \begin{bmatrix} \|\underline{u}_1\|^2, & \underline{u}_1^H \underline{u}_2 \\ \underline{u}_2^H \underline{u}_1, & \|\underline{u}_2\|^2 \end{bmatrix}^{-1} \left[\underline{u}_1, \ \underline{u}_2\right]^H \underline{\mathbf{a}} \\
&= \left[0, \ \underline{\mathbf{a}}^H \underline{u}_2\right] \begin{bmatrix} 1, & \cos\zeta \\ \cos\zeta, & 1 \end{bmatrix}^{-1} \begin{bmatrix} 0 \\ \underline{u}_2^H \underline{\mathbf{a}} \end{bmatrix} \\
&= \left[0, \ -\kappa_1^{-1}\right] \left(\frac{1}{\sin^2\zeta} \begin{bmatrix} 1, & -\cos\zeta \\ -\cos\zeta, & 1 \end{bmatrix} \right) \begin{bmatrix} 0 \\ -\kappa_1^{-1} \end{bmatrix} \\
&= \frac{1}{\sin^2\zeta} \left[+\kappa_1^{-1}\cos\zeta, \ -\kappa_1^{-1}\right] \begin{bmatrix} 0 \\ -\kappa_1^{-1} \end{bmatrix} \\
&= \left(\frac{\kappa_1^{-1}}{\sin\zeta} \right)^2
\end{aligned} \tag{8.93}$$

Therefore from Eqs. (8.91) and (8.93) we have

$$\begin{aligned}
\hat{\kappa}_1^{-1} &= \|\mathbb{P}_{\mathcal{H}_{12}} \underline{\mathbf{a}}(\breve{s})\| \\
&= \frac{\kappa_1^{-1}}{\sin\zeta}
\end{aligned} \tag{8.94}$$

which implies that

$$\hat{\kappa}_1 = \kappa_1 \sin\zeta \tag{8.95}$$

8.9.2 "Circular" and "Y" Arrays — Sensor Locations

Sensor number	"Circular" x	"Circular" y	"Y" x	"Y" y
1st	75.00	0.00	8.00	0.00
2nd	72.44	19.41	22.00	0.00
3rd	64.95	37.50	38.00	0.00
4th	53.03	53.03	57.00	0.00
5th	37.50	64.95	79.00	0.00
6th	19.41	72.44	105.00	0.00
7th	0.00	75.00	136.00	0.00
8th	−19.41	72.44	170.00	0.00
9th	−37.50	64.95	−4.00	6.93
10th	−53.03	53.03	−11.00	19.05
11th	−64.95	37.50	−19.00	32.91
12th	−72.44	19.41	−28.50	49.36
13th	−75.00	0.00	−39.50	68.42
14th	−72.44	−19.41	−52.50	90.93
15th	−64.95	−37.50	−68.00	117.78
16th	−53.03	−53.03	−85.00	147.22
17th	−37.50	−64.95	−4.00	−6.93
18th	−19.41	−72.44	−11.00	−19.05
19th	0.00	−75.00	−19.00	−32.91
20th	19.41	−72.44	−28.50	−49.36
21st	37.50	−64.95	−39.50	−68.42
22nd	53.03	−53.03	−52.50	−90.93
23rd	64.95	−37.50	−68.00	−117.78
24th	72.44	−19.41	−85.00	−147.22

8.9.3 Proof: CRB of Two Sources in Terms of κ_1

Let $\underline{u}_{11} \triangleq \underline{u}_1(s_1)$ and $\mathbb{A} \triangleq [\underline{\mathbf{a}}_1, \mathbb{A}_r]$ where $\mathbb{A}_r = [\underline{\mathbf{a}}_2, \ldots, \underline{\mathbf{a}}_M]$. Then the term $\underline{u}_{11}^H \, \mathbb{P}_{\mathbb{A}}^{\perp} \, \underline{u}_{11}$ of Eq. (8.38) can be rewritten as follows:

$$\begin{aligned}
\underline{u}_{11}^H \mathbb{P}_{\mathbb{A}}^{\perp} \underline{u}_{11} &= \underline{u}_{11}^H (\mathbb{P}_{\underline{\mathbf{a}}_1}^{\perp} - \mathbb{P}_{\underline{\mathbf{a}}_1}^{\perp} \mathbb{A}_r (\mathbb{A}_r^H \mathbb{P}_{\underline{\mathbf{a}}_1}^{\perp} \mathbb{A}_r)^{-1} \mathbb{A}_r^H \mathbb{P}_{\underline{\mathbf{a}}_1}^{\perp}) \underline{u}_{11} \\
&= \underline{u}_{11}^H \mathbb{P}_{\underline{\mathbf{a}}_1}^{\perp} \underline{u}_{11} - \underline{u}_{11}^H \mathbb{P}_{\underline{\mathbf{a}}_1}^{\perp} \mathbb{A}_r (\mathbb{A}_r^H \mathbb{P}_{\underline{\mathbf{a}}_1}^{\perp} \mathbb{A}_r)^{-1} \mathbb{A}_r^H \mathbb{P}_{\underline{\mathbf{a}}_1}^{\perp} \underline{u}_{11} \\
&= 1 - \underline{u}_{11}^H \mathbb{A}_r (\mathbb{A}_r^H \mathbb{P}_{\underline{\mathbf{a}}_1}^{\perp} \mathbb{A}_r)^{-1} \mathbb{A}_r^H \underline{u}_{11} \\
&= 1 - \underline{u}_{11}^H \mathbb{A}_r \left(\mathbb{A}_r^H \mathbb{A}_r - \frac{1}{N} \mathbb{A}_r^H \underline{\mathbf{a}}_1 \underline{\mathbf{a}}_1^H \mathbb{A}_r \right)^{-1} \mathbb{A}_r^H \underline{u}_{11} \qquad (8.96)
\end{aligned}$$

where $\mathbb{P}_{\underline{\mathbf{a}}_1}^{\perp} \underline{u}_{11} = \underline{u}_{11}$ has been used.

Substituting the above equation back into Eq. (8.38) and then expressing the columns of the matrix $\mathbb{A}_r$ as $\underline{\mathbf{a}}_i = \underline{\mathbf{a}}_1 + \Delta \underline{\mathbf{a}}_i$ (for $i = 2, \ldots, M$), the

CRB on the bearing p_1 can be expressed as a function of the array manifold parameters.

However, it is informative to evaluate the CRB for two emitters ($M = 2$) which are closely spaced at bearings p_1 and $p_2 = p_1 + \Delta p$, corresponding to arc lengths $s_1 = \breve{s} - \Delta s/2$ and $s_2 = \breve{s} + \Delta s/2$ respectively.

In this case the matrix $\mathbb{A}_r$ becomes the manifold vector $\underline{\mathbf{a}}_2$ and Eq. (8.96) is simplified as follows

$$\underline{u}_{11}^H \mathbb{P}_{\mathbb{A}}^{\perp} \underline{u}_{11} = 1 - \frac{N|\underline{u}_{11}^H \underline{\mathbf{a}}_2|^2}{N^2 - |\underline{\mathbf{a}}_1^H \underline{\mathbf{a}}_2|^2} \tag{8.97}$$

The physical proximity of the arriving signals allows the use of local differential geometry for the evaluation of Eq. (8.38) (or, of Eq. (8.36) for the general case of $M > 2$).

(a) Evaluation of the term $|\underline{u}_{11}^H \underline{\mathbf{a}}_2| = |\underline{u}_1(s_1)^H \underline{\mathbf{a}}(s_2)|$. Clearly:

$$\underline{u}_1(s_1)^H \underline{\mathbf{a}}(s_2) = \underline{u}_1(s_1)^H (\underline{\mathbf{a}}(s_1) + \Delta\underline{\mathbf{a}}) = \underline{u}_1(s_1)^H \Delta\underline{\mathbf{a}} \tag{8.98}$$

However, the manifold in the neighborhood of $\breve{s}$ may be interpreted as a circular arc of radius $1/\widehat{\kappa}_1(\breve{s})$ as depicted in Fig. 8.15. In such circumstances the inner product may be written as

$$\begin{aligned}\underline{u}_1(s_1)^H \Delta\underline{\mathbf{a}} &\simeq \|\Delta\underline{\mathbf{a}}\| \cos \Delta\psi \\ &\simeq \|\Delta\underline{\mathbf{a}}\| \sqrt{1 - \sin^2 \Delta\psi} \\ &\simeq \|\Delta\underline{\mathbf{a}}\| \sqrt{1 - \frac{1}{4}\|\Delta\underline{\mathbf{a}}\|^2 \widehat{\kappa}_1^2(\breve{s})}\end{aligned} \tag{8.99}$$

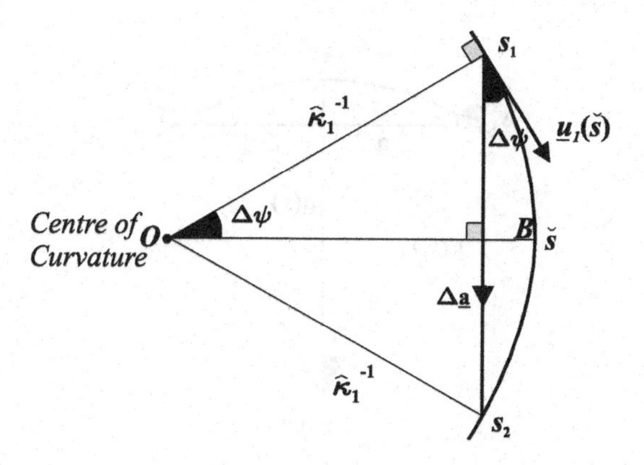

Fig. 8.15 Circular approximation of the manifold in the neighbourhood of $\breve{s}$.

or

$$\underline{u}_1(s_1)^H \underline{\mathbf{a}}(s_2) \simeq \|\Delta\underline{\mathbf{a}}\| \sqrt{1 - \frac{1}{4} \|\Delta\underline{\mathbf{a}}\|^2 \hat{\kappa}_1^2(\breve{s})} \tag{8.100}$$

(b) Evaluation of the term $\underline{\mathbf{a}}_1^H \underline{\mathbf{a}}_2 = \underline{\mathbf{a}}(s_1)^H \underline{\mathbf{a}}(s_2)$. Clearly:

$$\underline{\mathbf{a}}(s_1)^H \underline{\mathbf{a}}(s_2) = \underline{\mathbf{a}}(s_1)^H (\underline{\mathbf{a}}(s_1) + \Delta\underline{\mathbf{a}}) = N - \underline{\mathbf{a}}(s_1)^H \Delta\underline{\mathbf{a}} \tag{8.101}$$

The array manifold has constant norm and hence lies on the surface of a hypersphere. Consequently, for sufficiently small Δs, vectors $\underline{\mathbf{a}}(\breve{s})$ and $\Delta\underline{\mathbf{a}}$ are strictly orthogonal as indicated in Fig. 8.16. Using simple trigonometry:

$$\underline{\mathbf{a}}(s_1)^H \Delta\underline{\mathbf{a}} \simeq \sqrt{N}\|\Delta\underline{\mathbf{a}}\| \cos(90° + \gamma) \tag{8.102}$$

$$\simeq -\sqrt{N}\|\Delta\underline{\mathbf{a}}\| \sin(\gamma) \tag{8.103}$$

$$\simeq -\sqrt{N}\|\Delta\underline{\mathbf{a}}\| \frac{\|\Delta\underline{\mathbf{a}}\|}{2\sqrt{N}} = -\frac{1}{2}\|\Delta\underline{\mathbf{a}}\|^2 \tag{8.104}$$

Therefore

$$\underline{\mathbf{a}}(s_1)^H \underline{\mathbf{a}}(s_2) \simeq N - \tfrac{1}{2}\|\Delta\underline{\mathbf{a}}\|^2 \tag{8.105}$$

Substituting Eqs. (8.104) and (8.105) back into Eq. (8.97) and then using Eq. (8.38):

$$\mathrm{CRB}[p|\mathcal{A}] = \frac{1}{2(\mathrm{SNR}_1 \times L)\dot{s}(p_1)^2} \left(1 - \frac{N\|\Delta\underline{\mathbf{a}}\|^2 \left(1 - \frac{1}{4}\|\Delta\underline{\mathbf{a}}\|^2 \hat{\kappa}_1^2(\breve{s})\right)}{N^2 - (N - \frac{1}{2}\|\Delta\underline{\mathbf{a}}\|^2)^2}\right)^{-1}$$

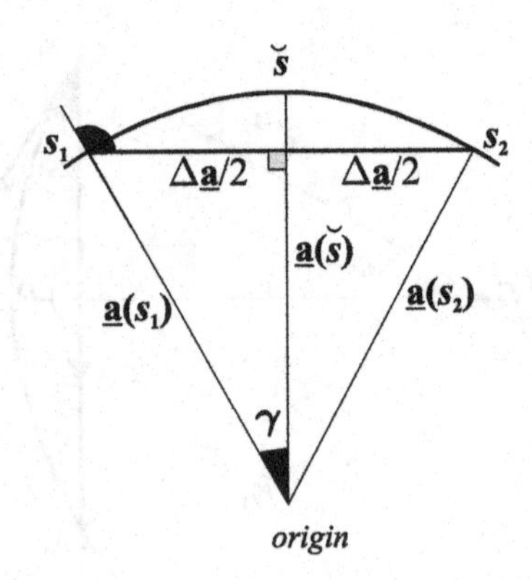

Fig. 8.16 For sufficiently small Δs, the vectors $\underline{\mathbf{a}}(\breve{s})$ and $\Delta\underline{\mathbf{a}}$ are strictly orthogonal.

$$= \frac{1}{2(\text{SNR}_1 \times L)\dot{s}(p_1)^2} \left(1 - \frac{N(1 - \frac{1}{4}\|\Delta\underline{\mathbf{a}}\|^2 \hat{\kappa}_1^2(\breve{s}))}{N - \frac{1}{4}\|\Delta\underline{\mathbf{a}}\|^2}\right)^{-1}$$

$$= \frac{1}{2(\text{SNR}_1 \times L)\dot{s}(p_1)^2} \left(\frac{4N - \|\Delta\underline{\mathbf{a}}\|^2}{\|\Delta\underline{\mathbf{a}}\|^2 (N\hat{\kappa}_1^2(\breve{s}) - 1}\right)$$

So finally:

$$\text{CRB}[p|\mathcal{A}] = \frac{1}{(\text{SNR}_1 \times L)} \frac{2}{\dot{s}(p_1)^2 \, (\Delta s)^2 \left(\hat{\kappa}_1^2(\breve{p}) - \frac{1}{N}\right)} \qquad (8.106)$$

where $|\Delta\underline{\mathbf{a}}| \simeq \Delta s = \Delta p \; \dot{s}(\breve{s})$ and it has been assumed that $4N \gg |\Delta\underline{\mathbf{a}}|^2$.

It is important to mention that a slight variation to the above expression may be obtained via a second-order Taylor expansion of $\underline{\mathrm{a}}(s_2)$ about s_1. However simulations indicate that Eq. (8.39) provides a more accurate approximation.

Bibliography

[1] R. Schmidt, *A Signal Subspace Approach to Multiple Emitter Location and Spectral Estimation.* PhD thesis, Stanford University USA, November 1981.

[2] U. Lipschutz, *Theory and Problems of Differential Geometry.* Shaum's Outline Series, McGraw Hill, 1969.

[3] M. D. Carmo, *Differential Geometry of Curves and Surfaces.* Prentice Hall Inc., New Jersey, 1976.

[4] H. Akaike, "A new look at the statistical model identification," *IEEE Trans. on Autom. Control,* Vol. 19, pp. 716–723, December 1974.

[5] M. Wax and T. Kailath, "Detection of signals by information theoretic criteria," *IEEE Trans. on Acoustics, Speech and Signal Processing,* Vol. 33, pp. 387–392, April 1985.

[6] M. Wax and I. Ziskind, "Detection of the number of coherent signals by the MDL principle," *IEEE Trans. on Acoustics, Speech and Signal Processing,* Vol. 33, pp. 1190–1196, August 1989.

[7] R. Schmidt, "Multiple emitter location and signal parameter estimation," *IEEE Trans. on Antennas and Propagation,* Vol. 34, pp. 276–280, March 1986.

[8] P. Karaminas and A. Manikas, "Superresolution broad null beamforming for co-channel interference cancellation in mobile radio networks," *IEEE Trans. on Vehicular Technology,* Vol. 49, pp. 689–697, May 2000.

[9] B. Steinberg, *Microwave Imaging with Large Antenna Arrays.* Wiley, 1983.

[10] I. Dacos and A. Manikas, "Estimating the manifold parameters of one-dimensional arrays of sensors," *J. Franklin Institute (Engineering & Applied Mathematics),* Vol. 332B, pp. 307–332, May 1995.

[11] A. Manikas, A. Sleiman and I. Dacos, "Manifold studies of nonlinear antenna array geometries," *IEEE Trans. on Signal Processing,* Vol. 49, pp. 497–506, March 2001.

[12] H. Karimi and A. Manikas, "Cone-angle parametrization of the array manifold in DF systems," *J. Franklin Institute (Engineering & Applied Mathematics),* Vol. 335B, No. 2, pp. 375–394, 1998.

[13] A. Manikas and C. Proukakis, "Modelling and estimation of ambiguities in linear arrays," *IEEE Trans. on Signal Processing*, Vol. 46, pp. 2166–2179, August 1998.

[14] A. Manikas, V. Lefkadites and C. Proukakis, "Investigative study of planar array ambiguities based on 'hyperhelical' parametrisation," *IEEE Trans. on Signal Processing*, Vol. 47, pp. 1532–1542, June 1999.

[15] H. R. Karimi and A. Manikas, "Manifold of a planar array and its effects on the accuracy of direction-finding systems," *IEE Proc. Radar, Sonar and Navigation*, Vol. 143, pp. 349–357, December 1996.

[16] A. Manikas, A. Alexiou and H. R. Karimi, "Comparison of the ultimate direction-finding capabilities of a number of planar array geometries," *IEE Proc. Radar, Sonar and Navigation*, Vol. 144, pp. 321–329, December 1997.

[17] H. Guggenheimer, *Differential Geometry.* McGraw-Hill (or Dover Edition), 1963 (1977).

[18] A. Sleiman and A. Manikas., "The impact of sensor positioning on the array manifold," *IEEE Trans. on Antennas and Propagation*, Vol. 51, pp. 2227–2237, September 2003.

[19] T. Willmore, *An Introduction to Differential Geometry.* Oxford University Press (UK), 1959.

[20] M. Spivak, *A Comprehensive Introduction to Differential Geometry.* Publish or Perish, Berkley, CA, 1979.

[21] L. Eisenhart, *Riemannian Geometry.* Princeton University Press, Princeton, NJ, 1949.

[22] M. Marcus and H. Minc, *A Survey of Matrix Theory and Matrix Inequalities.* Dover, 1992.

[23] H. Van Trees, *Detection, Estimation and Modulation Theory Part-1.* John Wiley and Sons, US, 1968.

[24] P. Stoica and A. Nehorai, "MUSIC, maximum likelihood and Cramer-Rao bound," *IEEE Trans. on Acoustics, Speech and Signal Processing*, Vol. 37, pp. 720–741, May 1989.

[25] Y. I. Abramovich, N. K. Spencer and A. Y. Gorokhov, "Detection-estimation of more uncorrelated gaussian sources than sensors in nonuniform linear antenna arrays—Part III: detection-estimation nonidentifiability," *IEEE Trans. on Signal Processing*, Vol. 51, pp. 2483–2494, October 2003.

Index

Printed in the United States
By Bookmasters